WITHDRAWN

AF722726

A10202 703697

QH
605
L87

CELLULAR MECHANISMS OF CHROMOSOME DISTRIBUTION

INTERNATIONAL REVIEW OF CYTOLOGY

Edited by

G. H. BOURNE

YERKES REGIONAL PRIMATE
RESEARCH CENTER
EMORY UNIVERSITY
ATLANTA, GEORGIA

J. F. DANIELLI

CENTER FOR THEORETICAL BIOLOGY
STATE UNIVERSITY OF NEW YORK
AT BUFFALO
BUFFALO, NEW YORK

Assistant Editor

K. W. JEON

CENTER FOR THEORETICAL BIOLOGY
STATE UNIVERSITY OF NEW YORK AT BUFFALO
BUFFALO, NEW YORK

SUPPLEMENT 1

Z. Hruban and M. Rechcigl, Jr., Microbodies and Related Particles: Morphology, Biochemistry, and Physiology

SUPPLEMENT 2

Peter Lukyx, Cellular Mechanisms of Chromosome Distribution

CELLULAR MECHANISMS OF CHROMOSOME DISTRIBUTION

PETER LUYKX

LABORATORY FOR QUANTITATIVE BIOLOGY
UNIVERSITY OF MIAMI
CORAL GABLES, FLORIDA

175612

1970

ACADEMIC PRESS New York and London

574.805
I61
Sup. 2

COPYRIGHT © 1970, BY ACADEMIC PRESS, INC.
ALL RIGHTS RESERVED
NO PART OF THIS BOOK MAY BE REPRODUCED IN ANY FORM,
BY PHOTOSTAT, MICROFILM, RETRIEVAL SYSTEM, OR ANY
OTHER MEANS, WITHOUT WRITTEN PERMISSION FROM
THE PUBLISHERS.

ACADEMIC PRESS, INC.
111 Fifth Avenue, New York, New York 10003

United Kingdom Edition published by
ACADEMIC PRESS, INC. (LONDON) LTD.
Berkeley Square House, London W1X 6BA

LIBRARY OF CONGRESS CATALOG CARD NUMBER: 52-5203

PRINTED IN THE UNITED STATES OF AMERICA

CONTENTS

EDITORS' FOREWORD

The organized movements of chromosomes in cells have for a long time fascinated cell biologists, and indeed many others. This monograph by Peter Luykx is a lucid account of the progress which has been made in understanding these phenomena. It will be of particular value for those interested in control of cellular processes, and for those concerned with establishing the molecular basis of cell behavior.

J. F. DANIELLI
G. H. BOURNE

CHAPTER I

INTRODUCTION

The persistence, or recurrence, of the particular morphological and physiological characteristics of living organisms is attributable partly to the relative constancy of their physical and chemical environment and partly to their innate capacity for self-maintenance and self-reproduction. It is a striking fact that individual living organisms do not simply maintain themselves indefinitely. Every individual cell or organism eventually dies, or else undergoes a process the result of which is the production of a larger number of cells or organisms than were present initially. Given the right conditions, the vast majority of these are fully capable of maintaining or recreating all of the morphological and physiological features characteristic of the particular kind of organism. The process by which new organisms are produced can be traced in every case to the division of individual cells, involving the transmission to each daughter cell of materials necessary for growth, differentiation, and further division.

The basic question prompting this review is: What are the basic cellular mechanisms responsible for the transmission of characteristic morphological and physiological features of organisms from one generation to the next? The observations briefly reviewed in this introduction support the propositions (1) that in general a precisely determined chromosome complement is an important prerequisite for basic cell function and for the development of multicellular organisms, and (2) that in most cells spindle poles and kinetochores acting together provide an important mechanism that insures the preservation of a specific chromosome complement in the production of two cells from one.

A. SIGNIFICANCE OF PRECISE CHROMOSOME DISTRIBUTION

The precise distribution of daughter chromosomes to daughter cells during cell division is a prerequisite for normal development and differentiation in multicellular organisms. Some precision of chromosome distribution is also normally required for the continuing growth and reproduction of individual cells, although here the requirements are slightly less exacting than for normal development in multicellular organisms.

These statements are based on the following observations and considerations.

1. Fate of Embryos with Duplications or Deficiencies

The importance of precise chromosome distribution for the reproduction of cells and the development of multicellular organisms can be estimated by observing what happens when chromosomes are distributed in a fashion less precise, i.e., in a way such that daughter cells have duplications or deficiencies for one or more chromosomes.

Deficiencies of even relatively small segments of chromosomes in *Drosophila melanogaster* are usually lethal to the embryo; duplications have less drastic effects, but even the smaller cytologically detectable duplications may cause morphological changes, and duplications of larger segments in most cases lead to reduced fertility and reduced viability (Li, 1927; see also Scriba, 1967, 1969). The embryonic death of many of the offspring of mice heterozygous for a reciprocal translocation has been traced to duplications and deficiencies of translocated chromosome segments transmitted to the zygote. This subject has been reviewed by Hadorn (1961).

The larger changes involved in duplications and deficiencies of whole chromosomes would be expected to have no less drastic effects on development. This was in fact first demonstrated by Boveri (1907), who showed that the developmental failure of isolated blastomeres obtained from dispermic sea urchin eggs cleaving in the same planes as normal eggs is a result of the abnormal distribution of chromosomes to four poles during the first cleavage division. The same conclusion can be drawn from the studies of Fankhauser and his co-workers on the relation between developmental abnormalities and chromosome complements in amphibians. A majority of salamander half-eggs, containing only sperm nuclei, fail to gastrulate successfully. This is not attributable to cytoplasmic deficiencies, because *diploid* egg fragments may develop normally; it is apparently a result of the deficient chromosome complements demonstrated in the cells of these half-eggs (Fankhauser, 1934a, b). Similarly, androgenetic *whole* eggs of *Triturus* fail to develop normally. A large proportion of even those eggs that appear to undergo normal cleavage fail to develop beyond gastrulation, probably because of the variable chromosome numbers shown to be present in the cells of these embryos (Kaylor, 1941). Abnormalities in development may be caused by extra chromosomes added to the diploid complement, as well as by absolute chromosome deficiencies. In the progenies of triploid axolotl females mated with diploid males, nearly all embryos have chromosome numbers between the diploid and triploid. These embryos develop abnormally and die relatively early. The small group of larvae showing less extensive abnormality and surviving for longer periods, however, is characterized by a high proportion of euploid (diploid and tetraploid) individuals (Fankhauser and Humphrey, 1950).

All these observations show that even if cytoplasmic materials are distributed normally during cleavage abnormal chromosome distribution and the consequent aneuploid chromosome constitutions are usually incompatible with the normal development of the embryo.

The occasional occurrence of nullisomic plants (e.g., Rana, 1967) might be explained by the presence of segments in other chromosomes, homologous to those in the missing chromosome pair. Results of analysis of meiotic pairing in haploid plants (see Kimber and Riley, 1963) make this a likely possibility.

2. Fate of Cells with Duplications or Deficiencies

Beyond its importance in development, is a balanced set of chromosomes also necessary for the basic processes of individual cell growth and division? The above observations do not bear on this question, for the death of aneuploid embryos may be caused by the failure of some specific gene product to be made by a particular group of cells or to its overproduction, rather than to the inviability of the cells themselves. The question may be divided into two parts. (1) What are the consequences, in terms of the reproductive survival of individual cells, of absolute deficiencies in chromosomal genes? (2) What are the consequences of alterations in the amount of chromosomal genetic material *not* involving absolute deficiencies, e.g., loss of one of two homologous chromosomes or addition of one or more extra chromosomes?

Two different approaches may be taken in attempting to assess the importance of a complete representation of chromosomal genes in cell growth and reproduction. One is to try to identify and then localize the cell components necessary for continuing survival and reproduction,

TABLE I

CHROMOSOMAL LOCATION OF STRUCTURAL GENES FOR SOME RNA AND PROTEIN SPECIES IMPORTANT IN BASIC CELL FUNCTION

	Location of structural gene	Organism	References
Ribonucleic acids			
rRNA	Nucleolar organizing region of X chromosome	*Drosophila*	Ritossa and Spiegelman, 1965; Ritossa *et al.*, 1965, 1966
	Chromosome	*Xenopus* (amphibian)	Birnstiel *et al.*, 1965; Wallace and Birnstiel, 1966
	Nucleolar organizing region of chromosome	Man	McConkey and Hopkins, 1964
tRNA	Chromosomes ("Minute" loci?)	*Drosophila*	Ritossa et al., 1965, 1966
Leucine tRNA	Nucleus	*Vicia faba* (bean)	Woods, 1965; Woods and Zubay, 1965
Enzymes or carriers concerned with energy metabolism			
Cytochrome *c*	Chromosome	*Saccharomyces* (yeast)	Sherman *et al.*, 1966, 1968
Malate dehydrogenase (mitochondrion)	Chromosome	*Neurospora*	Munkres *et al.*, 1965; Munkres and Richards, 1965
	Chromosome	Maize	Longo and Scandalios, 1969
	Chromosome	Man	Davidson and Cortner, 1967
Malate dehydrogenase (chloroplast)	Nucleus	*Acetabularia* (alga)	Schweiger *et al.*, 1967
Malate dehydrogenase (supernatant, NADP-dependent)	Autosome	Mouse	Henderson, 1966

β-Hydroxybutyric dehydrogenase	Nucleus	*Paramecium*	Tait, 1968
Phosphoglucomutase	Two or three chromosomal loci	Yeast	Bevan and Douglas, 1969
	Two autosomal loci	Mouse	Shows *et al.*, 1969
	Three autosomal loci	Man	Hopkinson and Harris, 1968
Glucose-6-phosphate isomerase	Autosome	Man	Detter *et al.*, 1968
	Linkage group I	Mouse	DeLorenzo and Ruddle, 1969
α-Glycerophosphate dehydrogenase	Chromosome II	*Drosophila*	Grell, 1967
NADP-Isocitrate dehydrogenase	Chromosome	Mouse	Henderson, 1965
Glucose-6-phosphate dehydrogenase	X Chromosome	*Drosophila*	Young *et al.*, 1964
	X Chromosome	Man	Boyer *et al.*, 1962; Nance, 1964; Yoshida, 1967
6-Phosphogluconate dehydrogenase	X Chromosome	*Drosophila*	Young, 1966
	(G-Group?) chromosome	Man	Fildes and Parr, 1963; Bowman *et al.*, 1966; Fialkow *et al.*, 1969
Enzymes concerned with nucleic acid synthesis			
Thymidine kinase	Group-E chromosome	Man	Migeon and Miller, 1968
Adenylate kinase	Autosome	Man	Raply *et al.*, 1967
Orotidine-5′-phosphate decarboxylase	Chromosome V	Yeast	Tauro *et al.*, 1968
Enzymes concerned with protein synthesis			
Isoleucine-tRNA synthetase	Chromosome II	Yeast	Hartwell and McLaughlin, 1968; McLaughlin and Hartwell, 1969
Methionine-tRNA synthetase	Chromosome	Yeast	McLaughlin and Hartwell, 1969
Polypeptide chain-initiating factor	Chromosome	Yeast	Hartwell and McLaughlin, 1969

in order to determine whether or not chromosomal genes are in fact an essential part of the cell as a self-reproducing system. The other approach is simply to observe the reproductive capacity of individual cells carrying absolute chromosomal deficiencies.

a. Location of "Essential" Structural Genes

Because the ability of a cell to grow and divide depends on its particular metabolic capabilities and on the culture conditions, it is not possible to define the minimal requirements for the reproductive survival of cells in absolute terms. But within the limits of our present knowledge, we can say that successful cell division means that each daughter cell must receive, in addition to a certain minimum of structural and enzymic equipment, the DNA templates for rRNA, for tRNA's, and for the mRNA's that code for the catalytic and structural proteins directly involved in energy metabolism, cell division, and the synthesis of lipids, proteins, and nucleic acids (see Morowitz, 1966). Table I lists a number of cases in which the structural genes for some of these "essential" cell components have been located on chromosomes. The evidence for a chromosomal location of these structural genes consists in the finding, in the case of proteins, that variations attributable to alterations in primary structure are inherited in Mendelian fashion or, in the case of RNA, that the RNA in question is synthesized in the nucleus or is capable of hybridizing specifically with chromosomal DNA. These observations, incomplete as they are, nevertheless sufficiently justify the need for the distribution to each daughter cell of at least one representative of each chromosome possessing one of these structural genes. They do not prove, however, that *all* chromosomes are absolutely essential for the survival of the cell as a reproducing entity. It is conceivable that the controlling elements for the cell components necessary for cell growth and division are in general not distributed over all chromosomes and that culture conditions could be found that would allow the reproductive survival of cells with absolute chromosomal deficiencies.

b. Effects of Duplications and Deficiencies

A cell deprived of its nucleus has at most only a limited capacity for growth and metabolic activity (see Brachet, 1955, 1957; Brachet and Chantrenne, 1956) and does not divide, indicating that at least within the limitations of current culture techniques the whole nucleus is an essential part of the cell as a self-reproducing system. At the other end of the scale of chromosome deficiencies are changes or losses of very small regions of a chromosome, perhaps in some cases involving only a single gene. If mutations or small deletions are in general lethal to the cell carrying them, then losses of whole chromosomes would also be expected to be lethal, making the distribution of a complete set of chromosomes to at least one daughter cell at division an obvious prerequisite for continuing cell reproduction.

David (1936) found that isolated fragments of chick embryos carrying a lethal factor (possibly a deletion) in the homozygous condition could be cultured *in vitro* or as chorioallantoic grafts for periods well beyond the time the embryo would have died. Although the persistent lower growth capacity of *heart cultures* from lethal embryos, as compared with those from normal embryos, suggests that some cell types might have defects at the cellular level, all other fragments in culture showed growth rates comparable to those of similar fragments from normal embryos. Similarly, in the mouse, not all mutations lethal to the embryo are lethal to every cell carrying the mutation. Ephrussi (1935) found that at least some regions of mouse embryos homozygous for the lethal gene *brachyury* proliferate *in vitro* as well as the corresponding regions from normal embryos. L. B. Russell concluded that "some X-linked mutations . . . known to be hemizygous lethal are not cell-lethal, for these mutations are, in fact, detected in functionally mosaic [i.e., with either one or the other of the two X chromosomes active in different groups of cells] heterozygous females. Moreover, many of them have arisen by somatic mutation in males where they are present in genetically (not functionally) hemizygous condition in part of the body" (Russell, 1964).

In *D. melanogaster* a number of small deficiencies in the X chromosome, lethal to the whole organism in homozygous condition and possibly representing slightly larger portions of the whole genome (of the order of 0.1–0.2%) than the examples just described, have been tested for their lethality to small groups of hypodermal cells (Demerec, 1934, 1936; Ephrussi, 1934). Somatic segregation in a cell of a female heterozygous for the deficiency would give one daughter cell homozygous for the deficiency; if it is capable of further growth and division and the X chromosomes are marked with appropriate mutant genes, this daughter cell would give rise to a patch of recognizable mutant tissue.

Such somatic segregation of deficiencies gave rise to viable groups of cells only when deficiencies for a small terminal region (*yellow, achaete, scute* loci) or for one particular interstitial segment (around *cut wings* locus) were involved (Demerec, 1934, 1936; Ephrussi, 1934); deficiencies involving 11 other known regions were cell-lethal (Demerec, 1934).

In a genetic and cytological study of the effects of homozygous deficiencies on cell viability in maize, McClintock (1938) found that the occasional somatic loss of a ring chromosome covering a homozygous deficiency for a segment adjacent to the centromere of chromosome V, representing about $\frac{1}{20}$ of the chromosome (about 0.5% of the whole genome), leads to the production of streaks of colorless tissue in the leaves. That such sectors are observed at all shows that the cells comprising them are viable; but these homozygous deficient tissues grow more slowly than the surrounding heterozygous tissue, as indicated by the frequently smaller size of the plant, tissue separations in homozygous sectors, reduced and narrowed leaves, and the small size of the colorless streaks. The deficient chromosome V in the haploid microspores leads to the formation of small, only partially filled pollen grains; the deficient male gametes are not transmitted to progeny. The deficient chromosome complement is readily transmitted through the female parent, however, showing that the deficient female gametes are functional. Similar studies on a larger deficiency in the same region of chromosome V, involving about one-seventh of the chromosome (about 1.5% of the whole genome), demonstrated that these homozygous deficient cells are not viable. Sections of root tips revealed rows of two, four, and eight enlarged, vacuolated cells—indicating that the cells can divide two or three times after the loss of the "protective" ring chromosome. Such cells apparently do not survive, for small patches of pycnotic and collapsed cells are also seen in the root tips, and colorless streaks of leaf tissue are not observed. The deficient chromosome complement results in empty pollen grains and is transmitted by neither male nor female parent.

From all of these observations it can be concluded that although some very small chromosome deletions may not be lethal to individual cells under suitable culture conditions, or surrounded by nondeficient tissue, under the same conditions most larger deletions are cell-lethal.

In line with the foregoing conclusion, cells with absolute deficiencies for whole chromosomes are observed only rarely. Cells with a subhaploid number of chromosomes were found by Morgenthaler (1948) in regions of originally haploid, irregularly cleaving salamander eggs transplanted to normal embryos, and may occasionally be seen in tumor cells or cells in culture (e.g., de Torok and White, 1960), but it is not known in these cases how long such cells can survive. In a few cases there is evidence that absolute deficiencies for whole chromosomes lead to failure of processes governing cell division. In an inbred strain of *Tetrahymena,* the deficient chromosome complements seen in the micronuclei of some cells, arising from abnormal distribution of chromosomes during vegetative growth, are apparently responsible for the abnormal micronuclear division observed in these cells during conjugation; in cells of the same strain, micronuclei with a more normal chromosome content behave more normally during conjugation (Nanney and Nagel, 1964). In the flowering plants *Tradescantia* and *Uvularia,* microspores deficient for one or more chromosomes usually do not reach division, while in the same cell populations other microspores without such deficiencies survive and undergo normal division (Conger, 1940; Barber, 1941).

Although aneuploid cells without absolute chromosome deficiencies are frequently viable, they often show rates of growth, division, or development that are slower than those of otherwise similar euploid cells. Hypo- and hyperdiploid microspores arising from abnormal meiotic divisions in autotetraploid *Tradescantia,* tomato, and rye reach the microspore division significantly later than do the diploid microspores in the same anthers (Gülcan and Sybenga, 1967). In a genetic study of diploid *Aspergillus nidulans,* Käfer (1961) demonstrated that a variety of aneuploid types—including trisomy ($2n + 1$) for seven of the eight linkage groups, $2n + 2$, $2n + 3$, $2n + 4$ and $n + 1$, $n + 2$, $n + 3$, $n + 4$—all grow more slowly than euploid (diploid or haploid) colonies. The observation that the frequency of aneuploid cells found in primary cultures of human leukocytes does not increase with the age of the donor (e.g., Bloom *et al.*, 1967) also suggests that aneuploid cells are at a disadvantage in competition with euploid cells *in vivo* (see also Ford, 1964; Taylor, 1968). It should be noted that under some conditions, in neoplasms and in established cell cultures, aneuploid cells may possess a selective advantage over the normal diploid cell (Hsu, 1961; Davidson, 1964; Ford, 1964; Harris, 1964; Mark, 1967), but it is obvious that even here the establishment of particular aneuploid cell types depends on a mechanism that will maintain a particular chromosome constitution.

The precise distribution of chromosomes during cell division, then, appears to be essential for the continuing reproduction of the daughter cells, since cells with absolute deficiencies for whole chromosomes or even parts of chromosomes do not survive and, indeed, to judge from the chromosomal location of genes involved in "basic" cell functions (Table I), would in general not be expected to. Since aneuploid cells in general are selected against in competition with euploid cells, it appears that even cells without absolute chromosome deficiencies require a balanced set of chromosomes for efficient growth and division.

With this conclusion in hand, one can list a number of selective advantages of a mechanism such as mitosis over one in which chromosomes are present as a large number of replicas distributed at random to two daughter cells at division (Gabriel, 1960; Swanson *et al.*, 1967): (1) A precise distribution mechanism insures that virtually every daughter cell receives a full complement of essential components. It is easy to show (see, for example, Gabriel, 1960) that the fraction of progeny cells expected to survive the random distribution of essential components decreases drastically with increasing numbers of such components and that under such a system successful reproduction of cells can be maintained only with the expense of a large number of copies of each essential component. (2) With a precise distribution mechanism, therefore, a smaller number of replicas is possible. This may be of selective advantage where nutritional or energy sources are limiting and would allow the segregation of advantageous mutations. (3) A precise distribution mechanism allows numbers of different essential components to be maintained in strict proportion to one another. This gene balance is, as we have seen, of some importance in the efficient regulation of basic cell function and is a prerequisite for the orderly development of multicellular organisms.

B. BASIC MECHANISM OF PRECISE CHROMOSOME DISTRIBUTION

Nearly all cells possess a mechanism for the precise distribution of chromosomes. [A possible exception may exist in the radiolarian *Aulacantha scolymantha*, whose more than 1000 chromosomes appear to be distributed from an equatorial region in a fashion less precise than is the case in ordinary cell division (Grell, 1953; Grell and Ruthmann, 1964).] A rough estimate of how precise the distribution of daughter chromosomes in mitosis is may be obtained from the observations given in Table II. These data do not include failures of the mechanisms

TABLE II

ESTIMATED FREQUENCIES OF MITOTIC NONDISJUNCTION[a]

	A	B	C	References
Tissue cells and primary cultures				
Aspergillus, diploid (trisomics)	17	—	0.02	Käfer, 1961
Tradescantia, pollen mother cells, second meiotic division	6	3/3339	0.03	Conger, 1940
Arbacia (sea urchin), blastomeres	44	1/92	0.05	German, 1966
Ambystoma (amphibian), first cleavage division	28	1/218,890	0.00003	Fankhauser and Humphrey, 1959
Herpestes (mongoose), male, various tissues	35	6/640	0.05	Fredga, 1967
Mouse, somatic tissues	40	8/1282	0.03	Ford, 1964
Mouse, T6 chromosomes	2	2/35,000	0.006	Ford, 1964
Mouse, bone marrow	40	0/248	<0.02	Mark, 1967
Mouse, pulmonary tissue	40	3/119	0.1	Mark, 1967
Man, leukocytes	46	118/18,105	0.028	Bloom *et al.*, 1967
Man, leukocytes	46	7/578	0.05	Hungerford *et al.*, 1969
Cell cultures and tumor cells				
Chinese hamster, ovarian cell culture (monosomic)	21	8/90	0.8	Kao and Puck, 1967
Man, fibroblasts	46	1/329	0.01	Hayflick and Moorhead, 1961
Mouse, Rous sarcoma	40	7/1071	0.03	Mark, 1967
Chinses hamster, Rous sarcoma	22	10/283	0.3	Kato, 1968
Man, tumor cells	46	5/222	0.1	Richart and Corfman, 1964

[a] Column A, number of chromosomes; column B, frequency of daughter cells with one chromosome more than the normal number; column C, frequency (%) of nondisjunctive events per chromosome per division (= column B ÷ column A × 100 × 2). In most cases cells with one chromosome *more* than the usual number have been taken as a more reliable estimate of the frequency of nondisjunction because chromosome *loss* by nondisjunction usually cannot be distinguished from chromosome loss caused by the preparation technique.

of chromosome distribution other than nondisjunction; e.g., multipolar spindles, spindle failure, chromosome loss. Similarly, cases of nondisjunction occurring in the first division of meiosis have been excluded because in this division failures of the distribution mechanism itself usually cannot be distinguished from failures in synapsis or crossing-over. It should be noted that the frequencies of aneuploid cells in these populations will accurately reflect the frequency of nondisjunction only if such cells divide just once; if the products of a cell division in which nondisjunction occurs are capable of undergoing more than one division subsequently, then the frequency of aneuploid cells in the population will be greater than the frequency of nondisjunction. Hence, these figures should be considered maximum estimates. In most cases, for a given chromosome, both daughter chromosomes are included in just one of the two daughter nuclei, rather than distributed to different daughter nuclei, only once in every 1,000 to 10,000 cell divisions. In spite of the different numbers of chromosomes represented, the range of frequencies of nondisjunction is surprisingly narrow. This suggests that the mechanisms of chromosome distribution have evolved to approximately the same degree of efficiency in widely different organisms. Under these circumstances the total frequency of nondisjunction, per division, would be higher in cells with relatively high chromosome numbers than in cells with low chromosome numbers, implying an upper limit to chromosome number in cells and organisms in which chromosome balance is an important prerequisite for continuing reproduction.

It is interesting to note that the data in Table II, meager though they are, suggest that the rate of nondisjunction is somewhat higher in tumor cells and cells in culture than in natural cell populations.

Two of the more conspicuously important elements involved in chromosome distribution in eukaryotic cells are the poles of the spindle and the kinetochores (centromeres) of the chromosomes. Normal cell division usually requires the presence of two kinetochores for each duplicated chromosome, and two spindle poles. In this sense the division apparatus is a doubly polarized system, the two sister kinetochores of each chromosome constituting one polarized element, the spindle the other. Other numbers of spindle poles or kinetochores are usually incompatible with precise chromosome distribution. Chromosomes without any kinetochores at all—e.g., "acentric" chromosome fragments induced by irradiation—do not behave normally on the spindle. While the two daughter halves of such fragments may separate at anaphase, they are distributed together to one pole or the other (Bajer, 1958a) or at least move to opposite poles at a frequency much lower than normal for intact chromosomes (Carlson, 1938). Single chromatids or undivided chromosomes, with just one kinetochore, appear to be distributed at random over the spindle (e.g., Newman, 1967; see review by Östergren, 1961). Since intact chromosomes in the same cells behave normally, the abnormal behavior of acentric fragments and of single chromatids cannot in general be ascribed to spindle malfunction. Similarly [except in the spermatocytes of certain insects (Metz *et al.*, 1926; Metz, 1933; Hughes-Schrader, 1935; Scott, 1936)], intact chromosomes interacting with a single spindle pole are not efficiently separated into two groups of daughter chromosomes (E. B. Wilson, 1901a, b; M. Boveri, 1903; Painter, 1918; LeFevre, 1907; Bělař and Huth, 1933; Schrader, 1947; Cleveland, 1958a). Multipolar divisions almost invariably produce aneuploid groups of daughter chromosomes at each pole (e.g., Boveri, 1907; Baltzer, 1908; Fankhauser, 1934b; Barthelmess, 1957; Bibring, 1962; Nur, 1963; but see Teplitz *et al.*, 1968; Pera and Schwarzacher, 1969, for some remarkable exceptions). In short, two spindle poles and two kinetochores for each chromosome are usually required for normal chromosome behavior and the maintenance of the normal chromosome complement during cell division.

C. POLES AND KINETOCHORES IN PROKARYOTIC CELLS?

Under suitable culture conditions 90–100% of the bacterial cells in a population are viable, i.e., capable of further growth and division (Valentine and Bradfield, 1954; Powell, 1958; Postgate *et al.*, 1961); this is on a par with the viability of mammalian cells in culture (Puck and Marcus, 1955; MacDonald and Bruce, 1968). Therefore, it may be assumed that there are definite mechanisms for ensuring the precise distribution of daughter chromosomes also in prokaryotic cells, perhaps involving elements comparable to the spindle poles and kinetochores of eukaryotic cells. This conclusion is supported by the observation that in some cases the radioactivity of thymine-H^3 supplied to dividing bacterial cells is subsequently distributed to daughter cells in a decidedly nonrandom fashion (Lark and Bird, 1965; Eberle and Lark, 1966; Lark, 1966; Chai and Lark, 1967; see also Cuzin and Jacob, 1965).

Most, perhaps all, prokaryotic cells are polarized, frequently growing in length with little or no increase in girth and dividing by transverse fission (e.g., *Escherichia coli*: Marr *et al.*, 1966) or, in the case of some spherical blue-green algae, dividing in planes successively at right angles (e.g., Allen and Stanier, 1968). In this sense the prokaryotic cell has poles. As suggested by Jacob *et al.* (1963), the polarized growth and division of the cell surface may be directly responsible for chromosome separation. In some species—*Streptococcus pyogenes, Bacillus cereus,* and *B. megaterium*—the addition of new cell wall material primarily to the equatorial region of the dividing cell (Cole, 1965; Jacob *et al.*, 1966) provides an efficient means by which two daughter chromosomes initially attached to the cell surface at the equator can be separated to the equatorial regions of the two daughter cells. Although in two species of *Salmonella* and in *E. coli* some new cell wall material appears to be added throughout the length of the growing cell (Cole, 1965; Beachey and Cole, 1966), certain elements of the cell wall in these cells also may undergo growth in a restricted equatorial zone. Schwarz *et al.* (1969) have recently shown that inhibition of murein synthesis in *E. coli* with low concentrations of penicillin leads to weakening and breakdown of the murein "sacculus" component of the wall in a very narrow equatorial zone. Thus, in several bacteria the localized equatorial growth of wall components serves to separate the poles of the cell during elongation and could account for the segregation of daughter chromosomes to daughter cells.

It is not entirely certain whether or not the bacterial chromosome possesses a specialized region directly responsible for its segregation, i.e., a kinetochore. That the bacterial chromosome is connected with the cell surface throughout most or all of the division cycle has considerable support from morphological studies. In many gram-positive bacteria this connection takes place through membranous or tubular infoldings of the plasma membrane known as mesosomes (Ryter and Jacob, 1964; Jacob *et al.*, 1966; Ryter, 1968). Sometimes the mesosome is morphologically differentiated in the region of contact with the chromatin material (Ellar *et al.*, 1967). Similar connections between chromosome and plasma membrane, while less conspicuous, may also be observed in gram-negative bacteria (Ryter and Jacob, 1966; Ryter, 1968; Schreil and Christensen, 1968; Pontefract *et al.*, 1969; see also Fuhs, 1969). The observations that in *Bacillus subtilis* smaller masses of chromatin material are attached to one mesosome, larger masses to two, and that extrusion of the mesosome from the interior of the cell by exposure to a hypertonic sucrose solution appears to drag the chromatin material to the cell surface, all support the idea that the attachment to mesosomes is functionally significant for the process of chromosome separation (Ryter and Jacob, 1964). Further observations on the spatial relationships between mesosomes and chromatin masses would be desirable, however, for a recent study (Highton, 1969) suggests that in log phase cells of *Bacillus licheniformis* 1.5–6 μ in length, with one to four chromatin masses per cell, there is usually only *one* mesosome per cell, which is associated with cross-wall formation and not in a position to separate daughter chromosomes.

The DNA at or very near the site of replication sediments at a rate different from that of the bulk of the cell's DNA and appears to be preferentially associated with detergent- and lipase-sensitive cell components (Hanawalt and Ray, 1964; Ganesan and Lederberg, 1965; Smith and Hanawalt, 1967). If these observations are taken to mean that the site of chromosome attachment to the cell surface represents the site of replication, then this attachment site must move along the length of the chromosome as replication proceeds and thus would not represent a "kinetochore" in the sense of a permanent, fixed region of the chromosome (Fig. 1a).

Other explanations of the characteristic sedimentation and associations of replicating regions are possible, however, and there is evidence suggesting that the moving replication site is *not* the same as the site responsible for separation of daughter chromosomes. The observations of Mason and Powelson (1956) on living cells of *E. coli* growing with a generation time of 30 minutes at 37°C reveal that active separation in each of the two chromatin masses of a cell, as indicated by the formation of lobulated and dumbbell-shaped "nuclear" areas, begins approximately 10 minutes before the completion of cell division. The studies of Cooper and Helmstetter (1968) indicate that at this time, under these growth conditions, each of the two chromosomes present is only about half replicated. These findings suggest that active chromosome separation can begin *before* replication is completed and imply that neither the sites of origin or termination of replication, nor the replication site itself, is directly involved in the separation of daughter chromosomes (Fig. 1b). These observations thus favor the existence of a specialized site responsible for chromosome segregation, located about halfway around the chromosome from the site of origin of replica-

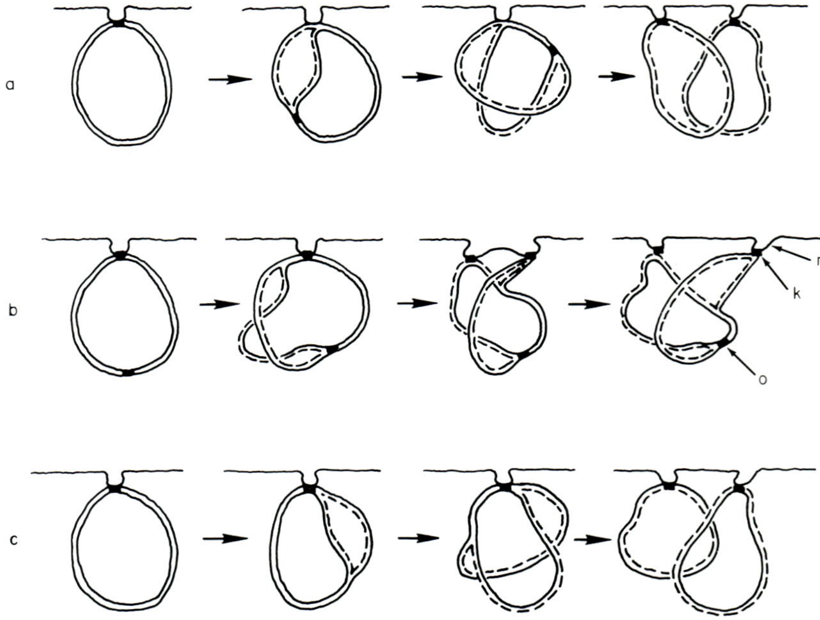

FIG. 1. Three models of the replication of a circular bacterial chromosome and separation of the daughter chromosomes. (a) Site of replication associated with cell surface (mesosome) throughout replication cycle; (b) specialized site (kinetochore), 180° from the origin, attached to mesosome; active separation of daughter chromosomes can begin when replication is half completed; (c) origin, acting as kinetochore, attached to mesosome throughout replication cycle; as in (a), active separation of daughter chromosomes cannot begin until replication has been completed. [*Legend* (see right end of row b)]: m, mesosome; k, kinetochore; o, origin of replication; dashed lines represent newly synthesized polynucleotide chains.

tion. Snyder and Young (1969), however, have recently found in stationary cells of *B. subtilis* (having no partially replicated chromosomes) that transforming markers associated with membrane fragments obtained by differential centrifugation are located at the *origin* and *terminus* of the chromosome; they suggest that the same may be true for growing cells. This observation again indicates a definite, fixed chromosomal site of attachment to the cell membrane, but one that would not allow separation of daughter chromosomes to begin until after replication has been completed (Fig. 1c).

More extensive studies on the temporal relations between chromosome replication and chromosome separation, and on the spatial relations between the chromosome and the cell surface, are required to resolve these apparent conflicts. More complicated models, such as those in which both the origin and the replication site are attached to the cell membrane throughout the cell cycle, or in which certain regions of the chromosome are attached to the cell surface for only a portion of the cell cycle, are not ruled out.

In short, the nonrandom distribution of chromosomes in the dividing bacterial cell requires some definite mechanism for chromosome segregation. This mechanism includes a functional equivalent to the polarized mitotic apparatus of eukaryotic cells probably involving the localized growth of components of the cell surface. The genetic, morphological, and biochemical evidence is inconclusive but is at least compatible with the idea that the attachment of some fixed chromosomal site, a kinetochore, to the cell membrane is responsible for the segregation of daughter chromosomes. Such a mechanism would be analogous to that apparently involved in chromosome separation during division in dinoflagellates (see Chapter III, Section C, 6, *d*).

CHAPTER II

SPINDLE POLES

The formation of the mitotic spindle in growing and dividing cells involves at least three different processes: (1) the synthesis of specific macromolecular spindle components from compounds of low molecular weight, (2) the assembly of the macromolecules into larger asymmetric structures such as microtubules, and (3) the orientation and association of the assembled structures to form the spindle. That these are distinct events is demonstrated by their separability. Even though dividing cells subjected to low temperature, colchicine, or other mitotic inhibitors show no signs of oriented spindle components or microtubules, the constituent macromolecules are probably present, for oriented spindle microtubules reappear very rapidly upon reversal of the inhibition (Inoué and Sato, 1967; Roth, 1967; Malawista *et al.*, 1968) and do so even in the presence of inhibitors of macromolecular synthesis (Inoué and Sato, 1967). The presence of a pool of preformed spindle precursors may also be inferred from the results of inhibition studies and studies on antigenic and colchicine-binding proteins of predivision cells (Taylor, 1959, 1965; Went, 1959, 1966a; Went and Mazia, 1959; Borisy and Taylor, 1967a, b; Gould, 1968; Rebhun and Sawada, 1969). Likewise, orientation of microtubules requires conditions in addition to those necessary for their formation, since in some cases microtubules are formed that do not appear to be oriented in preferred directions. This lack of definite orientation is characteristic of early division stages in *Haemanthus* endosperm (Bajer and Molè-Bajer, 1969) and in the ciliate *Campanella umbellaria* (Carasso and Favard, 1965). It also occurs in the polar regions of some dividing cells (P. Harris, 1965; Rebhun and Sander, 1967; Bajer, 1968a) and is found during spindle re-formation after treatment with low temperature (Roth, 1967) or chloral hydrate (Molè-Bajer, 1969) and in the cytoplasm of some nondividing cells (e.g., Gibbins *et al.*, 1969a).

The following is a summary of observations bearing on the questions of what determines the location of the spindle poles and what mechanisms are involved in the characteristic orientation of the microtubular components of the spindle. Related problems of spindle formation have been discussed by Mazia (1961) and by Went (1966a).

A. STRUCTURE

The region of the dividing cell toward which anaphase chromosomes move varies considerably in structure from one kind of cell to another. For convenience in description, spindle poles may be classified in three categories: poles organized around typical centrioles, poles organized around definite structures unlike typical centrioles, and poles organized around no definite or recognizable structure.

1. Spindle Poles Organized around Typical Centrioles

a. General Description.

At or near the center of the differentiated region at each end of the spindle in the majority of animal cells and in some lower plant cells lies a pair of centrioles, or occasionally a single centriole. The typical centriole is a small hollow cylinder 150–200 mμ in diameter and 300 mμ or more in length, whose wall is a complicated and precise longitudinal arrangement of nine triplet tubules embedded in a dense matrix. A detailed description of centriole structure may be found in the work of Stubblefield and Brinkley (1967) and de Harven (1968).

The *centrosome* includes the centrioles and associated osmiophilic materials ("pericentriolar halo," "dense granules," "satellites," and so on). The term *centrosphere* denotes the more-or-less differentiated region around the centrosome, frequently containing microtubules and membranous elements but excluding the larger cytoplasmic particles. The *aster* includes the centrosphere and the outlying radial arrangement of microtubules, membranes, rows of mitochondria, yolk and pigment particles, and so on. The use of these terms in this way is justified by correlations between recent electron microscope descriptions of mitosis and the observations of classical cytology, as indicated below.

A common feature, most conspicuous in the early stages of mitosis, is the presence of a pericentriolar halo of osmiophilic material (Fig. 2). This material may be diffusely distributed around one or both centrioles, or may be aggregated as projections from the surface of the centriole

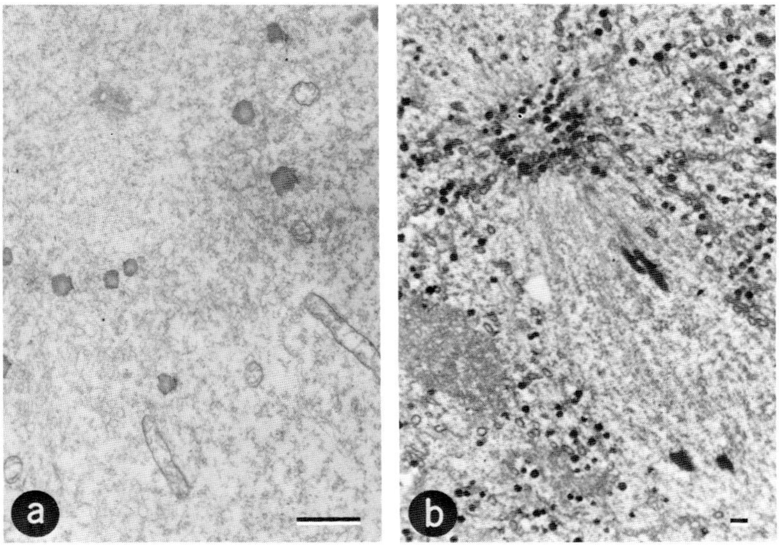

FIG. 2. Appearance of the polar region of the spindle in two stages of the first meiotic division in oocytes of the echiuroid worm *Urechis caupo*. (a) Early prometaphase. A portion of the centrosphere is shown in the upper left quadrant, delimited by an incomplete ring of lipid droplets; at the center of the centrosphere lies a pair of centrioles surrounded by electron-dense or osmiophilic material. (b) Midanaphase. A large number of lipid droplets and mitochondria have migrated centripetally from outlying regions of the aster to the boundary of the centrosphere. (Markers in lower right corners represent 1 μ.)

or as dense masses in the vicinity of the centriole (Robbins and Gonatas, 1964; Murray *et al.*, 1965; Phillips, 1966; Stubblefield and Brinkley, 1967; see especially Robbins *et al.*, 1968); it may be related to the pericentriolar satellites seen in some cells (Bernhard and de Harven, 1960; Bessis *et al.*, 1958; Szollosi, 1964; de-Thé, 1964; Gonatas and Robbins, 1965). The pericentriolar halo and the included centrioles together probably represent what was termed the "centrosome" in some of the older literature (compare Boveri, 1900, with Carasso and Favard, 1961, figures 21a and 78a, in *Ascaris megalocephala*; see also Bernhard and de Harven, 1960).

The variable appearance of the polar regions from one cell type to another, or even within one cell type during the course of division, is often due to the variable distribution of cytoplasmic components around

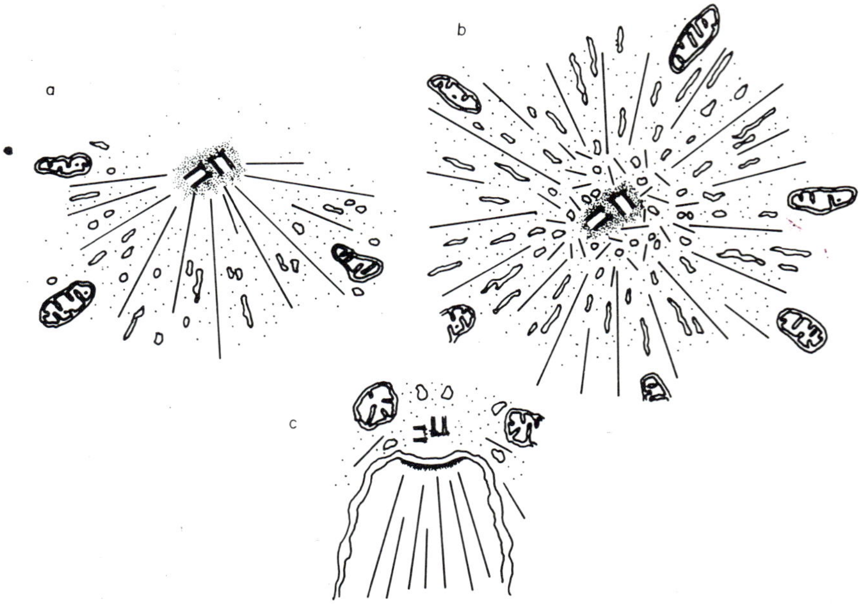

FIG. 3. Various appearances of spindle poles with centrioles. Microtubules are represented by straight lines. (a) Centrosphere consisting of concentric regions successively free of ribosomes, membranous elements, mitochondria, and other large cytoplasmic granules. (b) Centrosphere containing membranous elements and unoriented microtubules, excluding large cytoplasmic particles. (c) Intranuclear spindle; no centrosphere.

the centrioles. Early descriptions of spindle poles at the light microscope level have been summarized by E. B. Wilson (1925, pp. 672–684); they may be compared with the diagrams in Fig. 3).

Immediately outside the centrosome a region free of ribosomes and other particles is sometimes found (Fig. 3a). Slightly farther out many vesicles and other membranous elements occur; these are often more numerous near the spindle poles than elsewhere in the cell (Ito, 1960; P. Harris, 1962b; Manton, 1964b; Robbins and Gonatas, 1964; Roth *et al.*, 1966; Allenspach and Roth, 1967; Robbins *et al.*, 1968; Robbins and Jentzsch, 1969). Larger cytoplasmic particles, such as mitochondria, occur still farther out from the centrioles. Thus, around the centrioles in most cells there is a region free of these larger particles, but which contains many vesicles, microtubules, and ribosomes. This region is known as the centrosphere. Outside the centrosphere there is a radial arrangement of microtubules, elongated membranous elements, aligned mitochondria, and other cytoplasmic particles, probably representing the astral rays of light microscope descriptions (see Bessis *et al.*, 1958).

In cells with large asters (Fig. 3b), e.g., in the eggs of many marine invertebrates, the centrosphere is large and conspicuous. The microtubules in it do not show the definite radial or parallel orientation that exists outside the centrosphere (P. Harris, 1961, 1962a, 1965; Harris and Mazia, 1962; Rebhun and Sander, 1967). Although not a dividing cell, the melanophore of the fish *Fundulus* possesses a centrosphere and an aster remarkably similar to those of marine invertebrate eggs (Bikle *et al.*, 1966).

By contrast, other cell types possess poles in which centrospheres are less well defined and asters are inconspicuous. Even in these cases, however, certain cytoplasmic particles are often excluded from the vicinity of the centrioles. Microtubules are often radially oriented even in the immediate vicinity of the centrioles (see figures in Bernhard and de Harven, 1960; P. Harris, 1962b; Robbins and Gonatas, 1964; Krishan and Buck, 1965; Murray *et al.*, 1965; Roth *et al.*, 1966; Allenspach and Roth, 1967; Buck, 1967; Stubblefield and Brinkley, 1967; Robbins *et al.*, 1968).

In a few cases no centrosphere exists. Cytoplasmic particles such as mitochondria are found close to the centrioles (Fig. 3c). This is true in the fungi *Blastocladiella* (Lessie and Lovett, 1968), *Catenaria* (Ichida and Fuller, 1968), and *Saprolegnia* (Heath and Greenwood, 1968)—all of which have intranuclear spindles and extranuclear centrioles—and in secondary spermatocytes of the jellyfish *Phialidium* (Szollosi, 1964).

The variable distribution of cytoplasmic particles around the centrioles may reflect not only differences among cell types but also changes that occur during the course of division. Such changes have been described in the young sea urchin embryo (P. Harris, 1962b; Harris and Mazia, 1962) and in HeLa cells (Robbins and Gonatas, 1964; Robbins *et al.*, 1968; Robbins and Jentzsch, 1969). In the sea urchin embryo a tremendous increase in the number and concentration of vesicles around the centrioles accompanies the growth of the centrosphere and aster as the cell enters division. In HeLa cells a similar accumulation of vesicles around the centrioles occurs in late interphase and early prophase, excluding larger particles to form a small centrosphere. As the cell progresses from prophase to metaphase-anaphase and the centrosphere enlarges, both the vesicles and the larger particles (mitochondria, and so on) appear to be displaced outward; ribosomes may occur close to the centrioles but appear less numerous than slightly farther out. Telophase cells may show a small pericentriolar region free of ribosomes. After division has been completed, the centrosphere collapses; early interphase centrioles are closely surrounded by ribosomes, mitochondria, and other cytoplasmic particles.

b. Numbers and Orientations of Centrioles

Each spindle pole usually contains a pair of centrioles, but in the second meiotic division of oocytes and spermatocytes poles organized around single centrioles are common (e.g., von Korff, 1901; Payne, 1927; Johnson, 1931; Costello, 1961; Friedländer and Wahrman, 1966). When two centrioles are present at a pole, they often lie in a plane approximately perpendicular to the spindle axis (but exceptions are common, e.g., Mead, 1898; Huettner, 1933; Schrader, 1947; Murray *et al.*, 1965, Figs. 13, 14; Perkins and Amon, 1969, Fig. 15). The two centrioles usually lie at right angles to each other, sometimes parallel and perpendicular to the spindle axis (Krishan and Buck, 1965; Nagano, 1966; Phillips, 1967), sometimes with one centriole perpendicular to the spindle axis and the other at an angle (Murray *et al.*, 1965, Fig. 13), and sometimes with both centrioles at an angle to the spindle axis (e.g., Robbins and Gonatas, 1964, Figs. 23, 25; Turner, 1968, Fig. 9; Perkins and Amon, 1969). In cases in which there are only single centrioles at each pole, their axes may be parallel to each other and to the spindle axis (von Korff, 1901; Payne, 1927; Johnson, 1931), or they may be perpendicular to each other and to the spindle axis (Costello, 1961; Friedländer and Wahrman, 1966). Thus, the axis of the centriole itself appears, in general, to have no constant relation to the spindle axis. Nevertheless, in some specific cases—as that of the atypical centrioles in the spermatogonial divisions of the fungus gnat *Sciara* (Phillips, 1967)—there is apparently a strict relation between centriole axes and spindle axes during most of division. Costello (1961) has pointed out a correlation between centriole orientation and the orientation of the *spindle* at the *succeeding* division, thus raising the possibility that centriole orientation at one division determines spindle orientation at the next. These observations suggest that the variable relations between centriole axes and spindle axes described above are due to loss of centriole orientation or to reorientation once the spindle axis is established. A reorientation of centrioles has in fact been described at telophase in the division of the secondary spermatocytes of the hemipteran insect *Gelastocoris* (Payne, 1927), in spermatogonia of the gnat *Sciara* (Phillips, 1967), and in the antheridial cells of the alga *Nitella* (Turner, 1968).

The whole problem of the relation between centriole and spindle axes has not been systematically studied in any cell, however. It would be worthwhile to examine this relation in a cell type in which the orientation of the spindle at several successive divisions is predictable, e.g., in the cleavage divisions of many animal embryos.

The existence of spindle poles organized around single centrioles (see above) raises the question whether or not in the more usual case both members of a pair of centrioles are actually involved in the organization of a spindle pole. When microtubules in the immediate vicinity of a pair of centrioles are radially oriented, they sometimes appear to converge on only one of the two centrioles, e.g., Brinkley *et al.*, 1967, Fig. 11. [This centriole, Murray *et al.* (1965) suggest from their observations on rat thymic lymphocytes, is consistently oriented at right angles to the spindle axis; but in the chicken spermatocyte (Nagano, Fig. 27 in Fawcett, 1966) and in the fungus *Blastocladiella* (Lessie and Lovett, 1968), this centriole appears to be the one oriented *parallel* to the spindle axis.] A difference between the two centrioles of a pair with respect to the presence of pericentriolar osmiophilia has been claimed for some mammalian cells (Murray *et al.*, 1965; Robbins *et al.*, 1968), but to show this for certain would require serial sections. Structural differences between the two members of a centriole pair occur in *Sciara* spermatogonia (Phillips, 1967), which has the axis of the "primary" centriole consistently oriented parallel to the spindle axis. A functional difference between two members of a pair of centrioles is suggested also by Bibring's (1962) observation that the chromosomes of sea urchin eggs recovering from mercaptoethanol inhibition applied at prophase or prometaphase are distributed more unevenly to the four poles than would be expected on a random basis. In short, these indications of structural and functional differences between the two centrioles at a spindle pole suggest that only one of them is involved in the formation of the spindle pole; that spindle poles are organized around single centrioles is, in many secondary spermatocytes and oocytes, a matter of direct observation.

2. Spindle Poles Organized around Definite Structures Unlike Typical Centrioles

In some cells structures other than typical centrioles serve as focal points for the convergence of spindle fibers (see Fig. 4 for examples).

In the diatom *Surirella ovalis* the poles appear to be determined by granular spheres about 500 mμ in diameter (Drum and Pankratz, 1963; see Fig. 4a). In another diatom, *Lithodesmium undulatum*, the poles consist of two dense flat plates perpendicular to the spindle axis

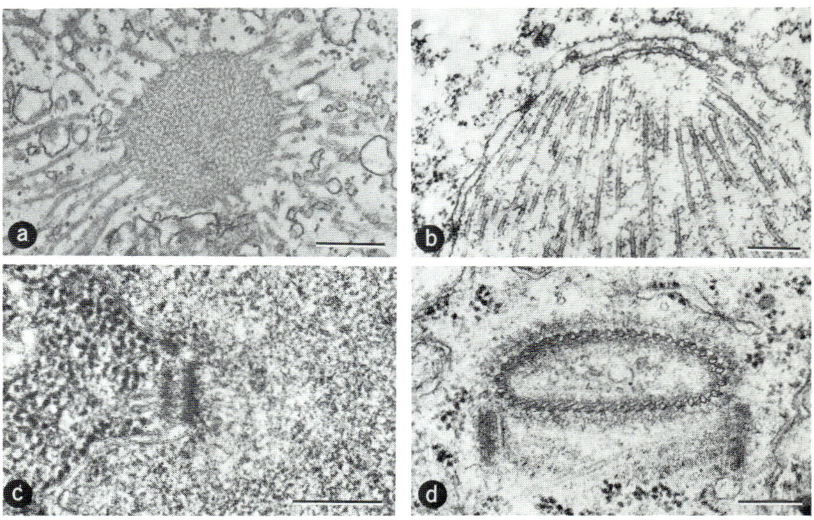

FIG. 4. Examples of the variety of structures found at the poles of the spindle in different organisms. (The marker represents 0.2 μ.) (a) Granular structure with radiating microtubules in the diatom *Surirella ovalis*. (b) Intranuclear spindle in the micronucleus of the ciliate *Nassula*; microtubules converge on a flattened "polar vesicle" inside the intact nuclear envelope. (c) Centriolar plaque, consisting of two dense disks separated by a lighter layer, associated with the nuclear envelope in *Saccharomyces* (yeast). The cytoplasm is to the left of the centriolar plaque and nuclear envelope, the nucleus to the right. (d) Atypical, giant centrioles in a spermatogonium of the gnat *Sciara*. The "primary" centriole has been sectioned transversely and shows, in this case, about 45 single tubules in its wall. Two sides of the wall of the "secondary" centriole, sectioned longitudinally, are seen below the primary centriole. A microtubule at an oblique angle can be seen in the lumen of the secondary centriole. [(a) From Drum and Pankratz, 1963; copyright 1963 by the American Association for the Advancement of Science; (b) from Tucker, 1967; (c) from Robinow and Marak, 1966; (d) from Phillips, 1967.]

(Manton *et al.*, 1969). In the dinoflagellate *Blastodinium*, Grell and Schwalbach (1965) found large masses of electron-dense material, about 1500 mμ in diameter, at the poles. A loosely organized mass of granular or amorphous material, about 600 mμ in diameter, occurs at one end of the asymmetrical spindle of the pollen mitosis in the orchid *Dactylorchis* (Heslop-Harrison, 1968). Division stages in the basidiomycetes *Coprinus lagopus*, *C. radiatus*, and *Armillaria mellea* show an ill-defined mass of fibrillar or granular material, about 300 mμ in diameter, at each pole (Lu, 1967; Lerbs and Thielke, 1969; Motta, 1967, 1969). Similar material is associated with spindle microtubules at the poles in the myxomycete *Physarum polycephalum* (Guttes *et al.*, 1968). In the micronucleus of the ciliate *Nassula*, a single flattened vesicle occurs at each end of the spindle, inside the intact nuclear envelope (Tucker, 1967; see Fig. 4b).

In the large complex flagellates symbiotic in the gut of roaches and termites, the spindle develops between "centriolar rods" or "centriolar bands" or "atractophores" composed of a dense fibrous material (Grimstone and Gibbons, 1966; Hollande and Valentin, 1967, 1968b,c; see also Cleveland, 1935, 1960). Spindle fibers insert directly on these structures (Cleveland, 1958b, in *Spirotrichonympha*) or on large dense spheres that are formed at their distal ends, as in *Barbulanympha* (compare Cleveland, 1953, and electron micrograph of J. Cook, fig. 45B in Mazia, 1961).

The poles of the intranuclear spindle in dividing yeast cells (*Saccharomyces*, an ascomycete) are marked by a double platelike structure, the "centriolar plaque," partially embedded in the intact nuclear envelope (Robinow and Marak, 1966; see Fig. 4c). Similar structures are also seen in other ascomycetes (Schrantz, 1967; Zickler, 1969). A somewhat similar local thickening of the intact nuclear envelope occurs at the poles of the intranuclear spindle in the microsporidian *Metchnikovella horvassei* (Vivier, 1965). The poles of the spindle in various stages of the life cycle of avian malarial parasites are composed of a diffuse electron-dense material associated with the nuclear envelope (Hepler *et al.*, 1966; Aikawa *et al.*, 1967; Terzakis *et al.*, 1967; Aikawa and Beaudoin, 1968).

These are all forms without flagellated stages and they lack typical centrioles at the poles of the spindle. It is interesting to compare them with the aquatic fungi *Saprolegnia*, *Allomyces*, *Blastocladiella*, and *Catenaria* (Heath and Greenwood, 1968; Robinow and Bakerspigel,

1965; Lessie and Lovett, 1968; Ichida and Fuller, 1968) and with the radiolarian studied by Hollande *et al.* (1969). In these organisms also, the spindle is formed within an intact nuclear envelope and the spindle microtubules terminate in a localized region of dense material on the inner side of the nuclear envelope. Immediately adjacent to these polar regions, however, outside the nucleus, typical centrioles are found. All of these organisms form flagellated zoospores.

These observations demonstrate that the complicated, nine-triplet-tubule structure of the typical centriole is not necessary for the formation of a normal, bipolar spindle. They suggest, rather, that the structure of the typical centriole is related to its role in the formation of cilia and flagella and not to the formation or orientation of spindle microtubules. This conclusion also finds support in the remarkable observations of Phillips (1966, 1967) on *Sciara* spermatogonia, in which very atypical centrioles—flattened cylindrical structures whose walls consists of about 70 single tubules in parallel (Fig. 4d)—appear at the poles of an otherwise typical spindle and are later involved in the formation of correspondingly atypical flagella.

3. Spindle Poles Not Associated with Any Definite Structure

The polar regions in some protozoa, some algae, slime molds, and the cells of higher plants (with the possible exception of the pollen mitosis in the orchid *Dactylorchis*, mentioned above) are not marked by any clearly recognizable structure that might play a special role in organizing the spindle. Characteristically, the microtubules of the spindle in such cells are oriented toward a broad, diffuse polar region and therefore converge only slightly at the poles. At some stages of division in amebas, they may not converge at all (Roth and Daniels, 1962; Daniels and Roth, 1964), so that the spindle is not spindle-shaped but cylindrical. Generally the radial arrangement of cytoplasmic particles, membranes, and microtubules at the poles is absent, i.e., the spindle is "anastral."

In dividing cells of higher plants, the polar regions of the spindle are often characterized by an abundance of membranous sheets and vesicles, first described by Porter and Machado (1960) in onion root tip cells and observed since then also in the root tip and coleoptile cells of wheat seedlings (Pickett-Heaps and Northcote, 1966), in *Haemanthus* endosperm (Harris and Bajer, 1965), and in pollen mother cells of *Lilium* (Dietrich, 1968) and *Tradescantia* (H. J. Wilson, 1967). Microtubules appear to be more disoriented in the polar regions than in the main body of the spindle and to become progressively more disoriented as division proceeds (Bajer, 1968a). Harris and Bajer (1965) describe the spindle pole in *Haemanthus* endosperm cells as a region from which mitochondria are excluded; but in dividing cells of the multicellular alga *Chara* (Pickett-Heaps, 1967) and in the giant ameba *Pelomyxa illinoisensis* (Daniels and Roth, 1964), the poles are regions where mitochondria accumulate. Such differences could be caused by differences in the plane of sectioning.

Although the study of division in these cells has generally been marked by a conspicuous failure to demonstrate any characteristic organelle at the poles of the spindle (e.g., Pickett-Heaps and Northcote, 1966), there are nevertheless a few indications that microtubules may terminate at specific sites in the polar regions. In some slime molds (Schuster, 1964; McManus and Roth, 1968) and in the micronucleus of a number of ciliates (Carasso and Favard, 1965; Jenkins, 1967), the spindle microtubules appear to terminate at a localized, although undifferentiated, region on the inner membrane of the intact nuclear envelope. In the slime mold *Physarum flavicomum* (Aldrich, 1967) and in the green algae *Chlamydomonas* (Johnson and Porter, 1968) and *Oedogonium* (Pickett-Heaps and Fowke, 1969), the situation is similar, except that the nuclear envelope is open at the poles. In *Chlamydomonas* the polar region is free of ribosomes and larger particles. Sheets of membrane in some sections through the polar regions give the impression of converging on one or a few focal points, and it has been suggested that these points serve as orienting centers for the developing spindle. The element of imagination appears to play a large part in such interpretations, however, especially since such arrangements can also be observed in the endoplasmic reticulum in other regions of the cell. In *Chara*, Pickett-Heaps (1967) found that the ends of microtubules sometimes appear to terminate in small spots of diffuse, electron-dense material in the cytoplasm and noted the presence of electron-dense fibrous material that could conceivably play a role in spindle formation in these cells.

B. THE SIGNIFICANCE OF VARIABILITY IN THE STRUCTURE OF SPINDLE POLES

All eukaryotic cells might use the same few basic devices for establishing the poles of the spindle. Variations in pole morphology would

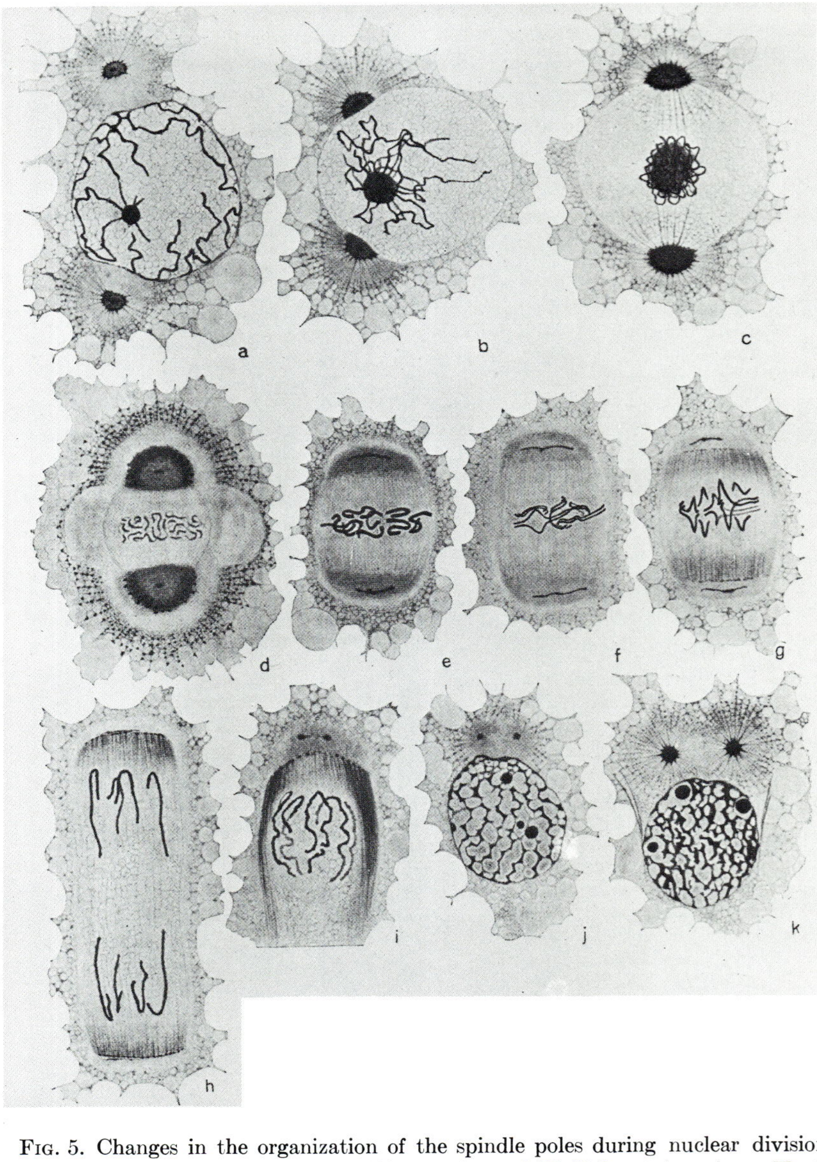

FIG. 5. Changes in the organization of the spindle poles during nuclear division in the gregarine *Monocystis magna*. (Observations of K. Belar, taken from Hartmann, 1933.)

then be only a consequence of differing relative contributions of these devices to spindle formation or might be the result of other differences in cell organization—differences in the number and kinds of inclusions, in the size and shape of the cells, in the stability of the nuclear envelope. This view can be justified by the following observations:

(1) Successive divisions in the same cell line may show differences in the morphology of their spindle poles, e.g., in the change from astral to anastral spindles during development of the sea urchin embryo (P. Harris, 1962b) and in cases in which cells with centrioles arise by division from cells without recognizable centrioles (see Mizukami and Gall, 1966; Turner, 1968, in plant spermatogenesis; Schuster, 1964, 1965; Aldrich, 1967, in the life cycle of slime molds). It seems unlikely that spindle formation would occur by fundamentally different mechanisms within a few cell generations in the same organism.

(2) Even within one division the morphology of the spindle poles may vary dramatically. A striking example is provided by Belar's observations on nuclear division in the gregarine *Monocystis magna*, in which the centrospheres and asters of the spindle at prophase and metaphase change during anaphase to broad polar regions without asters (Fig. 5). Along similar lines, Specht (1961) has described variations in the formation of the first maturation spindle in eggs of *Tubifex*; in some eggs the spindle is initially without asters, spindle fibers extending from the chromosomes toward diffuse, poorly defined polar areas; in other eggs the earliest stages of spindle formation are marked by large, conspicuous centrospheres and asters.

(3) Cells from widely different organisms may show similarities in the morphology of their spindle poles. For example, dividing cells of the parasitic protozoan *Metchnikovella*, the multicellular alga *Chara*, endosperm of the flowering plant *Haemanthus*, early cleavage stages of the sea urchin *Strongylocentrotus*, and HeLa cells all have concentrations of membranous elements at the poles of the spindle (see references above, Chapter II, Section A).

These observations suggest that in spite of the variability from cell to cell in the morphology of the spindle pole its formation in widely different cells may be brought about by fundamentally the same mechanisms. The observations presented in Chapter II, Section C suggest that in many cells spindle bipolarity is determined by (1) self-replicating centrioles or some material normally associated with the centrioles, acting together with (2) a mechanism based on interactions of spindle microtubules with each other or with some other preformed, oriented cell structure. In some cells one or the other of these two devices may predominate.

C. THE NATURE OF THE POLE DETERMINANT

Since the presence of a mitotic apparatus with more than two poles normally leads to irregular chromosome distribution and frequently cell death, the formation of a bipolar spindle at division is a prerequisite for the continuing reproduction of cells and organisms. What determines that at a particular place in the cell and at a particular time in the cell cycle a spindle pole appears? And why are there normally only two such sites in every dividing cell? Two problems inherent in these questions must be distinguished: the identity of the cell components immediately responsible for spindle formation, and the nature of the mechanism (or mechanisms) that ensures bipolarity.

Spindle formation will be viewed here primarily as a matter of the formation of oriented microtubules—the most asymmetric, conspicuous, and universal component of the division apparatus.

1. Evidence for Definite Pole-Determining Entities in the Cell

A widely accepted view is that there are definite organelles—pole determinants, or "mitotic centers"—directly responsible for spindle formation and capable of self-replication. There are two such organelles in every cell, and their replication is normally coordinated with cell reproduction and division, so that the presence of two functional pole determinants in every daughter cell is ensured.

That there are definite organelles responsible for the bipolarity of the spindle is supported by the observation that certain treatments give rise to a definite and sometimes predictable number of extra poles. Multiple or multipolar spindles are frequent in irradiated cell populations (Henshaw, 1940; Levis and Marin, 1963; Fetner and Porter, 1965) and in cells with endoreduplicated chromosomes or multiple nuclei (e.g., Roberts and Johnson, 1956; Schmid, 1966; Heneen *et al.*, 1967). Dividing sea urchin and sand dollar eggs recovering from inhibition by dinitrophenol, sodium azide, benzimidazole, or mercaptoethanol frequently form spindles with four poles instead of the usual two (Mazia *et al.*, 1960; Kojima, 1960; Went, 1966b); similar observations have been made on *Chaetopterus* (a marine annelid) eggs treated with urethan (Schuel, 1961). Inhibition of cytokinesis with caffeine or hexachlorocyclohexane in onion root tip cells results in the formation, at the next division, of two bipolar spindles in the same cell (Gonzalez, 1967; Giménez-Martín *et al.*, 1964; González-Fernández *et al.*, 1964).

All of these observations suggest that there are definite entities determining the location and formation of the poles—possibly in higher plant cells as well as in animal cells. In nondividing, enucleated cells of the early embryo of the sea urchin, the number of asters increases regularly with time, more or less in step with the duplication of those in dividing cells (Ziegler, 1898; Lorch, 1952). This shows that the appearance of new poles in this kind of cell is not directly dependent upon the nucleus, and is consistent with the notion that pole determinants are self-replicating.

2. Evidence That Centrioles are Pole Determinants

There is strong evidence that in the sea urchin egg these pole organizers are in fact the centrioles (Mazia *et al.*, 1960). Eggs treated with mercaptoethanol form quadripolar spindles on recovery and divide directly into four cells. The extra poles appearing on recovery from mercaptoethanol inhibition presumably do not arise as a result of an extra duplication of pole determinants because at the *subsequent* division only *half* the normal number of poles is formed (i.e., one instead of two) in each of the four cells. Therefore, it can be concluded that the effect of mercaptoethanol is only to allow the separation and independent functioning of preexisting pole determinants; that is, each pole determinant is normally a duplex structure. The time in the division cycle at which the pole determinant becomes double, determined as the earliest time at which mercaptoethanol can be added to give division from one cell to four, corresponds to the time at which the two cytologically visible poles separate in preparation for the coming division.

The similarity between these findings and electron microscope observations of centrioles is remarkable. There are *two* centrioles (a "diplosome") at each pole in the dividing sea urchin egg (P. Harris, 1961), as in most dividing animal cells. There is some evidence that only one centriole occurs at each pole of the quadripolar spindle in eggs recovering from mercaptoethanol inhibition (P. Harris, 1962a). The time of centriole duplication occurs around the time of pole separation in sea urchin eggs (Boveri, 1900) as in many other animal cells (Schreiner and Schreiner, 1905; Johnson, 1931; Minouchi, 1936; Murray *et al.*, 1965; Friedländer and Wahrman, 1966; Phillips, 1967; Robbins *et al.*, 1968). Thus, with respect to its normally duplex nature and the time of its duplication, the hypothetical pole determinant corresponds well with the pair of centrioles seen at each pole of the spindle.

This interpretation is in accord with electron microscope observations on the formation of the spindle in prophase in animal cells. In sea urchin eggs early stages in the formation of the aster involve the appearance of a dense mass of tubules and vesicles around the centrioles (P. Harris, 1961; Harris, and Mazia, 1962); in HeLa cells the first signs of spindle formation can be correlated with the appearance of an amorphous halo of dense material, dense bodies, vesicles, and microtubules around the centrioles (Robbins *et al.*, 1968).

Krishan and Buck (1965) and Stubblefield and Brinkley (1967) have interpreted some of their electron micrographs as showing direct connections of microtubules to centrioles, but more often the microtubules appear to end in the pericentriolar osmiophilic material or to attach to dense bodies or satellites in the vicinity of the centrioles (Robbins and Gonatas, 1964; Szollosi, 1964; Phillips, 1966; Allenspach and Roth, 1967; Robbins *et al.*, 1968). The convergence of microtubules at local regions containing an accumulation of dense material and membranous elements at the edge of the phragmoplast in *Haemanthus* endosperm (Bajer, 1968c) and in interphase cells of *Acanthamoeba* (Bowers and Korn, 1968) may be an expression of the same kind of activity. These observations, considered along with those on spindles with "atypical" centrioles (Chapter II, Section A,2) might mean that only the "dense material" is directly responsible for the formation and orientation of microtubules. Centrioles or other structures may act only indirectly in spindle formation by producing or accumulating the microtubule-orienting dense material around themselves.

3. Evidence for Self-Replication of Centrioles

The hypothesis that spindle bipolarity is determined by definite polar entities implies the existence of some mechanism for maintaining their number at the level of *two* in every cell about to enter division. This could be accomplished by self-replication of the pole determinant, coordinated with the reproduction of the cell as a whole.

There is some evidence that the centriole is a self-replicating structure. Cytochemical and radioautographic observations have indicated that DNA is closely associated with the basal bodies (structurally very similar to centrioles) in the ciliated protozoans *Tetrahymena* and *Paramecium* (Randall and Disbrey, 1965; Smith-Sonneborn and Plaut, 1967; see also Burton and Dusanic, 1968; Charret and André, 1968). Mechanisms of self-replication can be envisaged that do not require the presence of DNA (see discussions by Hoffman, 1965; Hanson, 1967; Phillips, 1967; Pyne, 1968), but the positive demonstration of DNA in centrioles or basal bodies would support the hypothesis of their self-replication and would be consistent with the identification of centrioles as the self-replicating pole determinants.

In spite of some cases in which new centrioles or basal bodies appear where none was evident before (Schuster, 1963; Dingle and Fulton, 1966; Outka and Kluss, 1967; Mizukami and Gall, 1966; van Assel and Brachet, 1966), in most actively dividing cells new centrioles appear to arise only adjacent to old ones of similar structure (Bernhard and de Harven, 1960; Gall, 1961; Murray *et al.*, 1965; Robbins *et al.*, 1968; Sorokin, 1968; see also Dippell, 1968; Allen, 1969). In embryos of *D. melanogaster* centrioles have been observed to persist as microscopically visible granules throughout the cell cycle (Huettner, 1933). The persistence of a centriole through several cell generations is clearly demonstrated in the early cleavage divisions of the trematode *Polystomum integerrimum*. According to the cytological studies of Minouchi (1936), one of the two poles in the first cleavage division contains a very long rod-shaped, easily visible centriole, while the other pole either lacks a centriole or contains a much smaller one not readily visible in the sections. The long rod-shaped centriole is recognizable at several succeeding cleavage divisions, at one pole in one of the blastomeres. It is interesting that the other pole, like the ones in all the other blastomeres, is without a recognizable centriole—demonstrating that duplication of the pole determinant does not necessarily involve exact replication of the whole centriole. The careful studies of Phillips (1967) on duplication of the atypical giant centrioles of *Sciara* spermatogonia have also shown that a recently formed centriole need not have the same structure as the older, adjacent one; but here the two poles of the spindle are alike, each having a pair of dissimilar centrioles. These observations are not necessarily inconsistent with the idea that the centriole is a self-replicating structure. Only a small portion of the centriole might replicate itself faithfully, allowing for some differences between the two daughter centrioles. It might also be that centrioles are self-replicating only in the sense that the microenvironment of a preexisting centriole is the only place in the cell suitable for the assembly of a new centriole.

4. Evidence for Additional Factors Determining the Number of Spindle Poles

There are several observations that do not agree with the view that self-replicating pole determinants are responsible for the formation of bipolar spindles. If the pole determinants are self-replicating, then it might be expected that a cell arising from the fusion of two parent cells would form four poles instead of two at the next division. In the earliest stages after the fusion of egg and sperm cells in fertilization, there may sometimes be indications of bipolar spindle formation associated with *each* gamete nucleus (Ferguson, 1901a; Conklin, 1904), but in the first division of the zygote a single bipolar spindle is the rule. Obviously, there may be special circumstances surrounding the events of fertilization. Even in the first division after the fusion of two somatic cells in culture, however, the expected multipolar spindle is not always found. The fact that the progeny of "hybrid" cells formed by the fusion of somatic cells in culture divide normally and contain complete sets of chromosomes from both parental cell types suggests that the original hybrid cell undergoes a normal *bipolar* division (Barski *et al.*, 1961; H. Harris *et al.*, 1966; Yerganian and Nell, 1966). Yamanaka and Okada (1968) have made direct observations on cell fusion and subsequent division; in their study most such cells formed one bipolar spindle and divided into two cells. Similarly, binucleate cells probably arising from the failure of cytokinesis often form a spindle with only two or three poles at the subsequent division. Fell and Hughes (1949) obtained film records of six binucleate cells in cultures from mouse tissues; each of the six formed a single bipolar spindle. Sisken and Kinoshita (1961) described several examples of binucleate cells forming tripolar spindles. In observations on 18 binucleate HeLa cells, Oftebro and Wolf (1968) found 9 with quadripolar spindles, 6 with tripolar spindles, 2 with bipolar spindles, and 1 case in which an initially quadripolar spindle was transformed to a tripolar spindle before anaphase.

Similar observations have been made on dividing multinucleate cells in higher plants. In the first meiotic division of barley pollen mother cells that fused or failed to complete cytokinesis, all the chromosomes line up on a single, large, bipolar spindle (Smith, 1942). Gonzalez (1967) described a sequence of events in caffeine-treated onion root tips in which nuclei originating from the fusion of nonsister telophase chromosome groups formed bipolar spindles at the next division.

In all these cases the normal duplication of pole determinants may be inhibited, the pole-forming activity of the extra pole determinants may be suppressed, or spindle bipolarity may not be strictly determined by self-replicating pole determinants alone.

Under some conditions *extra* poles may arise in dividing cells. In pollen mother cells of a number of higher plants, spindle disturbances in the first meiotic division lead to the appearance of extra spindles in the second meiotic divisions (Juel, 1897; Darlington and Thomas, 1937; Clark, 1940; Walters, 1958; Koopmans, 1958; Ting, 1966). Multipolar divisions can be induced in onion root tip cells by brief treatments with a variety of chemical agents (Barthelmess, 1957; Barthelmess and Einlechner, 1959; Barthelmess and Elkabarity, 1961), and in erythropoietic cells of the salamander *Molge vulgaris* by means of X-rays (Rondanelli *et al.*, 1967). In these cases multipolar spindles arise at times much shorter than one generation, and the spindles contain the normal number of chromosomes, so that it is unlikely that the appearance of extra poles is representative of the normal processes of cell reproduction.

To account for these observations it might be presumed that certain treatments induce premature or supernumerary duplications of the pole determinants or, alternatively, the pole determinant might be viewed as a compound structure capable of separation into subunits that retain the capacity to form extra poles. In many animal eggs, however, extra poles can arise independently of poles that might already be present. Matuszewski (1962) observed in eggs of the gall midge *Mikiola fagi* that unpaired meiotic chromosomes, eliminated from the spindle as part of the normal sequence of oogenesis, become associated with extra bipolar spindles, the number of such extra spindles apparently being a function of the number of groups of unpaired chromosomes. The extra asters ("cytasters") formed in many animal eggs in response to treatment by "activating" agents (see E. B. Wilson, 1925, for description and discussion) arise independently of previous spindles. That cytasters are functionally and structurally equivalent to the asters found at the poles of the normal spindle is indicated by the fact that they are capable of interacting with chromosomes and by the fact that they are often morphologically indistinguishable from spindle asters, even to the extent of having centrioles at their centers (Dirksen, 1961, in sea urchin eggs; Van Assel and Brachet, 1966, in amphibian eggs). In the amphibian eggs studied by Van Assel and Brachet (1966), heavy water induced the formation of many cytasters simultaneously; it is therefore unlikely that their appearance is a result of the duplication of the one or two pole determinants that might be present initially in the untreated cells. This possibility has been even more convincingly ruled out in the eggs of the nemertine worm *Cerebratulus*, in which the egg fragment remaining after cutting away the meiotic spindle can still form cytasters after activation (Yatsu, 1905). It must be concluded either (1) that there are additional pole determinants present, normally quiescent at least in the early stages of development but capable of being roused to pole-forming activity by the activating treatment (as suggested by Mazia, 1961, pp. 125–126), or (2) that cytasters can form entirely *de novo*—in effect, that spindle pole formation does not always require the presence of a definite entity derived from a preexisting pole.

In short, there is reasonably good evidence, even though indirect, for the view that in the sea urchin egg spindle poles are organized around definite, self-replicating entities: (1) the number of poles increases regularly in enucleated blastomeres; and (2) four poles instead of the usual two arise in eggs reversibly inhibited with certain chemical agents. This view probably holds also for a number of other animal eggs. As a general hypothesis, however, it fails to explain the variable numbers of poles that arise in artificially activated animal eggs, in various spontaneous and induced spindle abnormalities, and in cells formed by the fusion of two or more parent cells.

5. Absence of Centrioles or Functionally Equivalent Organelles in Some Cell Types

The hypothesis that the centriole is responsible for the polar organization of the spindle does not adequately account for observations on spindle structure and behavior in those cells in which neither centrioles nor other definite polar structures are found at the poles. It might be proposed, of course, that in all such cases "procentrioles" are present, difficult to recognize or detect because of their small size, poor staining characteristics, or organization in an unfamiliar form. Such a proposal would be difficult to disprove, but the following observations argue against it:

(1) Spindles in cells without centrioles may not respond to certain chemicals in the same way sea urchin and sand dollar eggs do. There have not, in fact, been any systematic studies on suitable cell populations, but in the ciliated protozoa *Stylonichia* and *Tetrahymena*, and in *Allium* root tips, there is so far no evidence that cells recovering from mercaptoethanol form quadripolar spindles (Dass and Sapra, 1966; Gavin and Frankel, 1966; Longevialle, 1968).

(2) Early stages of spindle formation in cells without centrioles may involve organization around several sites, resulting in a multipolar stage initially and changing to a bipolar organization only later in division. Some of the early observations on a multipolar stage in early spindle development in plants (Strasburger, 1895; Mottier, 1897; Davis, 1901; Osterhout, 1902) were later criticized by Dévisé (1922) as being attributable to fixation artifact, and indeed little evidence for such a stage has been obtained from polarization microscopy (Inoué, 1953; Inoué and Bajer, 1961). Nevertheless, it is doubtful that all such cases can be explained as artifacts. Ferguson (1901a,b) described spindle development in male and female gametophytes of pines, in which one of the two poles is unipolar from the beginning, while the other is initially multipolar, later converging to form a single second pole. Spindle formation in maize endosperm appears to begin with a stage in which fibrous material is oriented more or less radially around the early prophase nucleus, only later in prophase reorganizing to form typical "polar caps" (Duncan and Persidsky, 1958). Similar events seem to occur in the formation of the spindle from nuclear material in oocytes of the gall midge *Oligotrophus* (Matuszewski, 1966). In the first meiotic division of the egg of the scale insect *Pulvinaria*, the spindle is distinctly tripolar at prometaphase, but bipolar by metaphase (Nur, 1963). In the meiotic division of other coccid insects, the spindle also becomes bipolar only some time after the first appearance of spindle fibers (Hughes-Schrader, 1924, 1942, 1948b). Similarly, in the macronucleus of the ciliate *Campanella umbellaria*, bundles of microtubules appear to be randomly oriented with respect to one another in early division stages and become oriented more and more parallel with each other as division progresses (Carasso and Favard, 1965). Convergence of microtubules at several sites is characteristic of early stages in the

re-formation of the spindle on recovery from cold treatment in the giant ameba *Chaos carolinensis* (Roth, 1967).

Recently, the occasional occurrence of a definite multipolar stage in spindle formation has been documented in living *Haemanthus* endosperm cells by means of differential interference microscopy (Bajer and Molè-Bajer, 1969). Just before the nuclear envelope breaks down at the end of prophase, the developing spindle may have anywhere from two to five distinct poles. By metaphase the spindle is invariably bipolar. Electron microscopy of the same material shows that microtubules are more irregularly arranged in the earliest stages of spindle formation but gradually become oriented in parallel as the spindle becomes bipolar.

If a structure comparable to a centriole were responsible for spindle formation in all these cells, an early multipolar stage would not be expected, for the development of the typical centriolar spindle appears to be bipolar from the beginning.

(3) There are significant differences between spindles with centrioles and those without, with respect to the distribution of birefringence and the effect of localized ultraviolet irradiation. In metaphase spindles with centrioles birefringence is generally strongest at the poles; in spindles without centrioles birefringence is strongest near the chromosomes at the equator and very weak in the polar regions (Inoué, 1964). In a metaphase spindle with centrioles, UV microbeam irradiation of a small area between the equator and one pole results in loss of birefringence only in the irradiated region (Forer, 1965, 1966), while microbeam irradiation of the equivalent area in a spindle without centrioles results in loss of birefringence not only in the irradiated area but also in the region between that area and the pole (Inoué, 1964).

All these observations argue for the view that, in cases where centrioles are not seen at the poles, the orientation of spindle microtubules is caused by factors other than a reorganized or inconspicuous centriolar material.

6. Other Mechanisms of Spindle Formation and Microtubule Orientation

Even in cells *with centrioles,* mechanisms other than the activity of the centrioles may also participate in the orientation of spindle material and the establishment of bipolarity. In crane fly spermatocytes, according to observations of living cells by polarizing microscopy (Dietz,

1963), spindle material not immediately adjacent to centrioles may sometimes become oriented first in relation to individual chromosomes; only later do the individual "chromosomal spindles" line up in parallel and merge to form a single bipolar spindle. In some spermatocytes highly flattened between slide and cover slip, one of the spindle poles may be located in a region far from the centrioles (whose positions may be inferred in the living cell from the locations of the bases of the flagella). Such a pole is without an aster and may not always converge to a sharply defined point, but chromosome orientation and movement toward it is normal (Dietz, 1959, 1963, 1969; see also Dietz's photographs of the living cells in Went, 1966a). Cleavage may occur between the astral and anastral poles, giving rise to one daughter cell without any centrioles. At the next division this cell may form a bipolar spindle in which chromosome behavior is entirely normal (Dietz, 1966).

In the course of normal division in the unicellular alga *Chlamydomonas reinhardi*, the flagellar basal bodies appear to be involved in the formation and orientation of microtubules associated with cleavage of the whole cell but are not associated with the development of the spindle. The poles of the spindle are located at large openings in the otherwise intact nuclear envelope and are associated with areas free of ribosomes and other cytoplasmic particles; apparently, however, they lie some distance from the centriolelike basal bodies (Johnson and Porter, 1968). A somewhat similar situation has been described for the development of the maturation spindle in fertilized eggs of the rotifer *Asplanchna priodonta*; here, the fully formed spindle has a large sperm-aster at one pole, but it appears to arise first as an anastral spindle completely separate from the sperm-aster (Storch, 1924). In the complex flagellates *Trichonympha* and *Pseudotrichonympha*, typical centrioles or basal bodies occur at the bases of the flagella and in the "cap" or "rostral tube"; the mitotic spindle develops not in association with these centrioles, however, but from the distal ends of rods or bands of electron-dense fibrous material (Grimstone and Gibbons, 1966).

That factors besides centriolar activity may be involved in the establishment of spindle bipolarity is also emphasized by the observation that occasionally two or more of the poles of already established multipolar spindles may fuse to form one pole. The fusion of two of the four poles of a binucleate HeLa cell to give a tripolar spindle at metaphase was observed by Oftebro and Wolf (1968). Pole fusion was also described by Campbell and Inoué (1965) in *Pectinaria* oocytes recovering from UV microbeam irradiation. In abnormal spermatocytes of the snail *Paludina* (*Viviparus*), a cluster of centrioles arises at each pole early in meiosis I (see Gall, 1961). At the end of the first meiotic division, the centrioles are scattered around the periphery of the cell. If the centrioles were the sole factors determining spindle polarity, it would have been predicted that a highly multipolar spindle would appear at the next division; but Meves (1903) described a sequence of events in the secondary spermatocytes in which the centrioles reaggregated at opposite poles of the cell to form a bipolar spindle again.

What other mechanisms, aside from the activity of centrioles or related structures, could bring about the orientation of microtubules and the formation of a bipolar spindle?

One possibility is that in many cells there is a permanent cytoplasmic organization that permits spindle microtubules to form only in certain preferred orientations. One can imagine a persistent "framework" upon which the mitotic spindle is laid down at each successive division. The bipolarity of the spindle would be traceable not to specific self-replicating orienting centers or pole determinants, but to the maintenance of this framework, this permanent polarized organization of the cell (as in most if not all prokaryotic cells), possibly by morphogenetic gradients or by mechanisms akin to those that maintain specific patterns of organization in the cortex of ciliates (see Sonneborn, 1963; Beisson and Sonneborn, 1965; Nanney, 1968). This hypothesis is most plausible for cells in which the axis of the mitotic spindle has a definite relation to a consistent, specific organization of the cell as a whole, e.g., root meristematic cells, many ciliated and flagellated protozoa. It seems less applicable to cells that appear from microscopic observation to be less rigidly organized, in which there is no obvious consistent relation between the positions of the poles of the developing spindle and preexisting cell organization, e.g., in cells in culture, and in animal eggs, in which spindle axes at successive cleavage divisions or even at different stages within the course of one division have different orientations with respect to the egg surface. Even in these cases, however, there could conceivably be a specific pattern of cytoplasmic organization that is not apparent from the distribution of visible cytoplasmic components (see E. B. Wilson, 1925, Ch. XIV; Raven, 1966, pp. 30–55; see also Crick and Hughes, 1950).

Another possibility is that formation of the bipolar spindle in cells without centrioles is governed primarily by interactions among the spindle components themselves, especially by interactions between spindle microtubules. Ordered systems of tubular and fibrous elements, structurally similar to the spindles and asters of dividing cells, can be seen as "liquid crystals," or "tactoids," in solutions of tobacco mosaic virus (Bawden *et al.*, 1936; Bernal, 1940; Bernal and Fankuchen, 1941) and in solutions of sickle-cell hemoglobin (J. Harris, 1950; Murayama, 1966). Like the spindles of dividing cells, the ordered structures of sickle-cell hemoglobin are reversibly disrupted by high pressure and by low temperature (Murayama, 1966; see Pease, 1946; Zimmerman and Marsland, 1964; Inoué, 1964; Roth, 1967). Aggregates of microtubular filaments oriented in parallel, without apparent relation to specific orienting structures, are also found in a variety of cells (Ledbetter and Porter, 1963; Jordan, 1964; Newcomb and Bonnett, 1965; Tilney and Porter, 1965; Cronshaw and Esau, 1967; Holmes and Choppin, 1968).

These examples show that parallel aggregations of microtubules can occur in the absence of specific "orienting centers" or definite pole-determining structures, and that such aggregations may be governed by physical variables such as temperature and pressure in the same way the mitotic spindle is.

The view that the mitotic spindle is a kind of tactoid has also been discussed by Swann (1952) and by Schrader (1953).

The parallel orientation of spindle microtubules by their interactions with each other, or with more-or-less stable preexisting cell structure, could explain the formation of spindles with only two poles even in cells having twice the usual number of every other cell component, e.g., in cell fusion. It could also explain the formation of spindles and asters in regions of the cytoplasm far removed from the "primary" spindle, e.g., cytasters, supernumerary poles and spindles occurring as spontaneous formations or in response to various chemical and physical treatments. Consistent with this interpretation is the observation that adjacent parallel microtubules in spindles (Jenkins, 1967; H. J. Wilson, 1969; Bajer and Molè-Bajer, 1969) and in other microtubular systems (Grimstone and Cleveland, 1965; Kessel, 1966; McIntosh and Porter, 1967; Bannister and Tatchell, 1968; Peters *et al.*, 1968; Hepler and Jackson, 1968; Tilney and Byers, 1969) sometimes show cross-connecting filaments or side arms that could be involved in the orientation of the microtubules. Even in cells with centrioles, some mechanism involving interactions of spindle components must often be invoked to account for the orientation of spindle microtubules and the establishment of polarity. It is likely that in cells without visible microtubule-orienting, pole-determining structures some such mechanism is entirely responsible for the formation of bipolar spindles.

D. SUMMARY

Correlations between hypothetical pole determinants or mitotic centers on the one hand, and centrioles on the other, with respect to numbers, positions, and time of duplication, strongly support the idea that in many animal cells the centrioles play a decisive role in determining the locations of the poles of the spindle. It is not known whether the centriole directs the synthesis of spindle macromolecules, is involved in their assembly to form larger structural components of the spindle, determines the orientation of spindle components, or whether it plays only an indirect role in these processes.

Ultrastructural studies on fungi with and without a flagellated stage in the life cycle, and on cells with atypical centrioles, indicate that the characteristic nine-triplet-tubule structure of the typical centriole is related to the formation of cilia and flagella and is not essential for spindle formation. The association of microtubules with pericentriolar osmiophilic material or with centriolar satellites, or with similar dense material in a number of cells without typical centrioles, suggests that this dense material is somehow more directly involved in the assembly and orientation of spindle microtubules than is the typical centriole.

Correlations between the behavior of the hypothetical pole determinant and that of the centriole do not extend to all experimental situations. The number of poles appearing at division in cells resulting from cell fusion or failure of cytokinesis is not predictable on the hypothesis of a self-replicating pole-determining entity such as the centriole. One must postulate the existence of factors regulating the duplication or activity of pole determinants or admit that other mechanisms responsible for spindle bipolarity exist, even in cells with centrioles. Furthermore, under some conditions in both plant and animal cells, spindles or asters may appear at sites apparently unrelated to the positions of existing or previous spindle poles.

These observations, in conjunction with those indicating that the details of spindle formation and the response to experimental manipulation may differ between cells with centrioles and cells without centrioles, suggest that mechanisms other than a definite, self-replicating pole determinant help to ensure spindle bipolarity and that these other mechanisms are the predominant ones in cells without centrioles. The parallel orientation of microtubular structures seen in solutions of tobacco mosaic virus and of sickle-cell hemoglobin suggests that mutual interactions of spindle components, especially of spindle microtubules, may be one way in which bipolar spindles are formed in the absence of definite pole determinants. The highly organized cortex and cytoplasm of ciliates and hypermastigote flagellates suggests another possibility, that spindle components may interact with preexisting, self-perpetuating, oriented cell structure to form, invariably, a bipolar spindle.

The problem of whether or not the mechanisms responsible for microtubule orientation and spindle formation in cells without centrioles are fundamentally different from those operating in cells with centrioles deserves more attention. For example, are the "multipolar" stages in the microsporocytes of higher plant cells described around the turn of the century really the result of fixation artifact? In cells that normally divide with centrioles do microtubules appear prior to their association with centrioles? Can all such cells, experimentally deprived of the centrioles at one or both poles, form bipolar spindles at the next division, as in crane fly spermatocytes? The question of differences in spindle formation between cells with centrioles and those without could also be approached by looking for similarities and differences between them with respect to the effects of agents known to affect spindle formation and structure in specific ways. For example, can cells *without centrioles* be induced to form quadripolar spindles by means of treatment with mercaptoethanol, as in sea urchin eggs (Mazia *et al.*, 1960)? Can the mitotic apparatus of cells *with centrioles* be disrupted to form individual chromosomal spindles by means of treatment with chloral hydrate, as is apparently the case in *Haemanthus* endosperm (Molè-Bajer, 1967)? Informative in this connection would be more-detailed, comparative studies on the conditions controlling the number of spindle poles arising after cell fusion or failure of cytokinesis in plant and animal cells.

CHAPTER III

KINETOCHORES

The kinetochore (centromere) is the differentiated chromosomal structure to which spindle fibers attach during mitosis or meiosis. Except for the rare events involving inversion or translocation of a chromosome segment, the position of the kinetochore is constant for a particular chromosome, maintained during the gradual prophase condensation of the chromosome, and from one cell generation to the next. It may be located anywhere along the length of the chromosome (see Levan *et al.*, 1964). The kinetochore region is often one of the most conspicuous features of linear differentiation along the length of the chromosome, a specific functional unit at a particular location in the chromosome, analogous in that respect to specific loops of lampbrush chromosomes, or to specific bands or puffs of polytene chromosomes.

A. EVIDENCE THAT THE KINETOCHORE IS RESPONSIBLE FOR CHROMOSOME MOVEMENT IN CELL DIVISION

The kinetochore is responsible for normal chromosome movement and distribution to daughter cells during cell division. This is indicated by the observations that in normal cells the kinetochore usually leads the way in active chromosome movements during all stages of division, prometaphase as well as anaphase (Bajer and Molè-Bajer, 1956; Bajer, 1958c). Kinetochore regions of chromosomes sometimes appear to be "stretched" when the chromosomes first come into contact with spindle fibers, around the time of nuclear envelope breakdown (Bajer and Molè-Bajer, 1962, 1963; Dietz, 1969). Throughout the stages of active chromosome movement, spindle fibers are associated with the chromosome at the site of the kinetochore (Inoué and Bajer, 1961); this is true not only during anaphase, when daughter chromosomes separate and move to the poles of the spindle, but also during the earlier stage of prometaphase, when chromosomes move to the equator (P. Harris, 1965; Luykx, 1965a; Bajer, 1968b). Nicklas and Staehly (1967) have demonstrated in their micromanipulation studies on living grasshopper spermatocytes that the distance of the kinetochores from the poles does not change even though the chromosome is stretched with a microneedle to several times its original length; the kinetochores are more firmly attached to the spindle than are other parts of the chromosome. The same conclusion can be drawn from the observation that in fixed *Haemanthus* endosperm cells the shrinkage of chromosomes that occurs during dehydration takes place without a change in the position of the kinetochore (Jensen and Bajer, 1969).

Chromosome fragments may be induced by a variety of agents, such as X-irradiation, certain chemicals, and so on; fragments may also occur as a result of spontaneous chromosome breakage, or as a result of meiotic crossing-over in inversion heterozygotes. Fragments not containing the kinetochore region behave abnormally at division, either being eliminated from the spindle or lagging in the equatorial region at anaphase (e.g., Carlson, 1938; Bajer, 1958a; Taylor *et al.*, 1962). Clear examples of the difference in behavior between whole chromosomes and acentric fragments can be seen after crossing-over between normal and inverted chromosome segments, where frequently only one chromosome is involved and the one or two acentric fragments produced can be identified unequivocally [see, for example, the beautiful photomicrographs of Brandham (1969)].

That the arms of the chromosome do not usually play any essential role in chromosome movement is indicated by the normal mitotic behavior of minute centric fragments (Lima-de-Faria, 1955a; John and Hewitt, 1968). Kodani and Stern (1946) reported finding a small chromosome among the offspring of X-irradiated *Drosophila* males that was too small to be seen in mitotic cells, but apparently behaved normally during mitosis because it could be seen as a small polytene chromosome in salivary gland cells and its transmission could be followed genetically.

The effects of irradiating localized regions of chromosomes with a UV microbeam are generally consistent with the conclusion that the kinetochore is responsible for normal chromosome movement. Irradiation of chromosome regions other than the kinetochore has no effect on chromosome movement, except for stickiness preventing separation at the irradiated spot and leading to a chromatin bridge as kinetochores move toward the poles at anaphase. In *Triturus* heart cells in culture, chromosomes irradiated in the kinetochore region at prometaphase fail to reach the equator of the spindle (Bloom *et al.*, 1955). Irradiation of one of the two kinetochore regions of a bivalent in grasshopper spermatocytes at metaphase causes the bivalent to move in the direction of the pole toward which its intact kinetochore region is oriented (Izutsu, 1961). Such bivalents remain off the metaphase plate or reorient toward the opposite pole but do not undergo normal anaphase disjunction. The failure to behave normally at anaphase is not attributable to the irradiation of nearby spindle fibers; irradiation of a small area just on the poleward side of a kinetochore causes the bivalent to shift toward the opposite pole, but it recovers its position on the metaphase plate and undergoes normal anaphase disjunction. Irradiation of kinetochores at anaphase prevents the irradiated chromosomes from being included in the daughter nuclei. In *Haemanthus* endosperm cells irradiation of anaphse kinetochores often causes reorientation and movement of the chromosome toward the equator of the spindle, although irradiation of kinetochores at prometaphase appears to have no effect on congression to the metaphase plate (Bajer and Molè-Bajer, 1961).

B. DOES THE KINETOCHORE HAVE OTHER FUNCTIONS?

In addition to its role in chromosome movement during division, the kinetochore region has been implicated in (1) the initiation of synapsis and localization of chiasmata, (2) the terminalization of chiasmata, (3) chromosome condensation or coiling, (4) the association of sister chromatids until the beginning of anaphase, and (5) the formation or assembly of the chromosomal spindle fibers themselves.

1. Initiation of Synapsis

Darlington (1965) has listed a number of organisms in which the pairing of homologous chromosomes at meiosis appears to begin in the region of the kinetochore. The restriction of chiasmata to this region in some organisms could be explained by assuming that in these organisms the pairing process operates only for a limited time (Darlington, 1940), not long enough for pairing along the whole length of the chromosome. Crossing-over would naturally take place only within the regions already paired when the time limit occurs, i.e., those regions nearest the kinetochore. Other explanations of nonrandom distribution of chiasmata at metaphase are also possible, e.g., complete pairing followed by localized crossing-over. It is also not certain that the kinetochore regions do in fact possess special pairing properties. In some organisms pairing may begin at the distal ends of chromosomes, or at other sites far from the kinetochore regions (see Moens, 1969). In organisms in which the kinetochore regions are the first to pair, they might do so only because they are closer to each other at the beginning of synapsis, having converged on a spindle pole during anaphase of the last premeiotic mitosis (see discussion and references in Wagenaar, 1969). Furthermore, it should be remembered that in translocation heterozygotes and in polyploids synapsis and chiasma formation can occur between distal homologous chromosome segments connected to nonhomologous or nonpairing proximal segments, so that the pairing of homologous kinetochore regions cannot be considered a prerequisite for the pairing of distal regions.

2. Terminalization of Chiasmata

In some organisms the chiasmata seen in meiotic prophase (diplonema) are located anywhere along the length of the chromosome, but by metaphase are located predominantly or exclusively at the ends of the chromosomes. The movement of chiasmata toward the ends of the chromosomes is known as *terminalization* (see Darlington, 1965) and, as far as can be determined from fixed cell preparations, it occurs more or less continuously from diplonema through diakinesis to prometaphase or metaphase. It is often assumed that the kinetochore region plays a special role in this process. Observations on metacentric chromosomes are not particularly informative in this regard because in such chromosomes movement of a chiasma toward the nearest end is necessarily movement away from the kinetochore region. In telocentric chromosomes it might appear that chiasmata do indeed move away from the kinetochore region, for at metaphase terminal chiasmata are almost always located at the other end of the chromosome. It is important to distinguish, however, between terminalization occurring during prophase and terminalization occurring during prometaphase. In the grasshoppers *Stenobothrus parallelus* (Darlington and Dark, 1932) and *Melanoplus femur-rubrum* (Hearne and Huskins, 1935), the sudden increase between diakinesis and metaphase I in the proportion of chiasmata that are terminal suggests that the interaction of kinetochores with the spindle can indeed affect the position of chiasmata.

Poleward pulling on opposed kinetochores during prometaphase could move chiasmata away from the kinetochores and thus could account for the distal location of single terminal chiasmata in telocentric chromosomes at metaphase. There is in fact no convincing evidence to show that the terminalization occurring before spindle formation and breakdown of the nuclear envelope is always a movement of chiasmata away from the kinetochore region. The observations of Hearne and Huskins (1935) show that during prophase in *M. femur-rubrum* spermatocytes some chiasmata terminalize by moving *toward* the terminal kinetochores. Their data suggest that during prometaphase these chiasmata may then become interstitial again by moving away from the kinetochore region as the kinetochores are stretched toward the poles of the developing spindle. These observations are consistent with hypotheses that attribute the terminalization of chiasmata during prophase to general properties of homologous chromosomes without recourse to special properties of the kinetochore region (see also John and Lewis, 1965, for discussion).

3. Initiation of Chromosome Condensation

The possibility that the kinetochore region plays some role in initiating chromosome condensation as the cell enters mitosis rests on two kinds of observation. First, in many organisms the degree of condensation in prophase is greater in the regions of the kinetochores than in more distal regions of the arms (e.g., Lima-de-Faria, 1956). Second, in cells whose chromosomes are large enough, light microscope observations often reveal a reversal of coiling direction, from a right-handed to a left-handed coil, across the kinetochore region (Darlington and Vosa, 1963). Such reversals may also occur across chiasmata and secondary constrictions, however, and may be occasioned by any relatively uncoiled region of the chromosome, not by specific properties of the kinetochore itself. It should also be remembered that the physical association of a kinetochore with any particular chromosome segment is not necessary for mitotic condensation, since acentric fragments induced during interphase undergo apparently normal condensation. It seems likely that characteristic coiling or condensation patterns depend primarily on local properties of the chromosome segments involved. The possibility that the kinetochore region exerts an influence over the degree of condensation at any point along the chromosome could be tested by careful observations of the effect of structural alterations, such as deletions or inversions, on the normal pattern of chromosome condensation.

4. Association of Sister Chromatids

The general region of the kinetochore has special properties affecting the association of the two sister chromatids. The association of sister chromatids cannot be attributed to the kinetochore region alone, because in many cells at metaphase sister chromatids are closely apposed all along their length. Even the two chromatids of acentric fragments remain associated until anaphase (Bajer, 1958a; Bajer and Molè-Bajer, 1963). Nevertheless, with respect to chromatid pairing, the kinetochore region is somehow different from other regions of the chromosome. When sister chromatids part at the beginning of normal anaphase or after treatment with colchicine, the kinetochore regions are often the last to separate (see Lima-de-Faria and Bose, 1962; Holm and Bajer, 1966). At the beginning of anaphase in the first division of meiosis, sister chromatids often disjoin everywhere except in the kinetochore regions; the kinetochore regions remain associated until anaphase of the second division. Similarly, in the gall midge *Dasyneura crataegi*, at a certain stage in larval development the polytene chromosomes of salivary gland cells normally dissociate into a large number of strands; the kinetochore region is generally the last region to dissociate (S. A. Henderson, 1967).

The adhesiveness of sister chromatids in the kinetochore region is due neither to the kinetochore itself nor to the underlying chromatin, but to special regions immediately adjacent to the kinetochore, the proximal regions of the arms. In squash preparations fixed with alcohol and acetic acid (e.g., Lima-de-Faria, 1955b; J. Y. Wilson, 1956; Ohnuki, 1968) and in whole-chromosome mounts prepared for electron microscopy (DuPraw, 1966; Abuelo and Moore, 1969), sister chromatids appear to be in contact on either side of the kinetochore region but not within the kinetochore region itself, so that a "hole" appears between sister kinetochores. In living *Haemanthus* endosperm cells, a space between sister kinetochores is visible at metaphase and can be seen very clearly during poleward stretching of the kinetochore region in prometaphase (Bajer and Molè-Bajer, 1962; Bajer, 1966a; see also Fig. 7).

[Sometimes, however, sister chromatids appear to be in close contact even within the kinetochore region, and no such "hole" is visible (e.g.,

Brinkley and Stubblefield, 1966; Lubs and Blitman, 1967). Metaphase chromosomes both with and without the kinetochore "hole" can be seen in one and the same cell (Ohnuki, 1968). The cause of this variation is not clear; conceivably it could reflect real differences between chromosomes attributable to asynchronous chromosome condensation or chromatid separation, or to asynchronous development of associations with spindle fibers, as suggested by Jokelainen (1967).]

Is the delayed separation of sister chromatids on either side of the kinetochore the result of a delay in the completion of DNA replication in these regions? At first sight this might seem to be a likely possibility, because the heterochromatin adjacent to the kinetochore, the "proximal heterochromatin" characteristic of the chromosomes in many organisms, often replicates its DNA after DNA replication in most other regions of the chromosome has been completed (e.g., Bianchi and de Bianchi, 1965; Keyl and Hägele, 1966). If the association of sister chromatids at metaphase is due to the presence of small regions of unreplicated DNA, then this DNA should replicate just before anaphase separation. Comings (1966) incubated cultures of human fibroblasts for a short time in a medium containing tritiated thymidine of high specific activity, but even after prolonged exposure of the radioautographs failed to find any labeling over the kinetochore regions of early anaphase chromosomes. The observation is inconclusive, however, because, as Stone *et al.* (1965) have shown for *Tetrahymena*, tritiated thymidine is taken up into cellular DNA precursor pools only during the main period of DNA synthesis, in interphase. Inhibitors of DNA synthesis applied to lily microsporocytes at late prophase appear to have no effect on the separation of chromatids during the subsequent meiotic divisions (Ito *et al.*, 1967). Although not conclusive, these observations suggest that chromosome components other than unreplicated stretches of DNA are responsible for the association of sister chromatids adjacent to the kinetochores. One could also argue for the same conclusion on the general grounds that in many mitotic cells sister chromatids normally remain closely associated all along their length long after the interphase doubling of DNA.

5. Formation of Spindle Fibers

The birefringence seen in living cells at metaphase and anaphase is not uniformly distributed throughout the spindle; much of it is concentrated in streaks running between the poles and the kinetochore

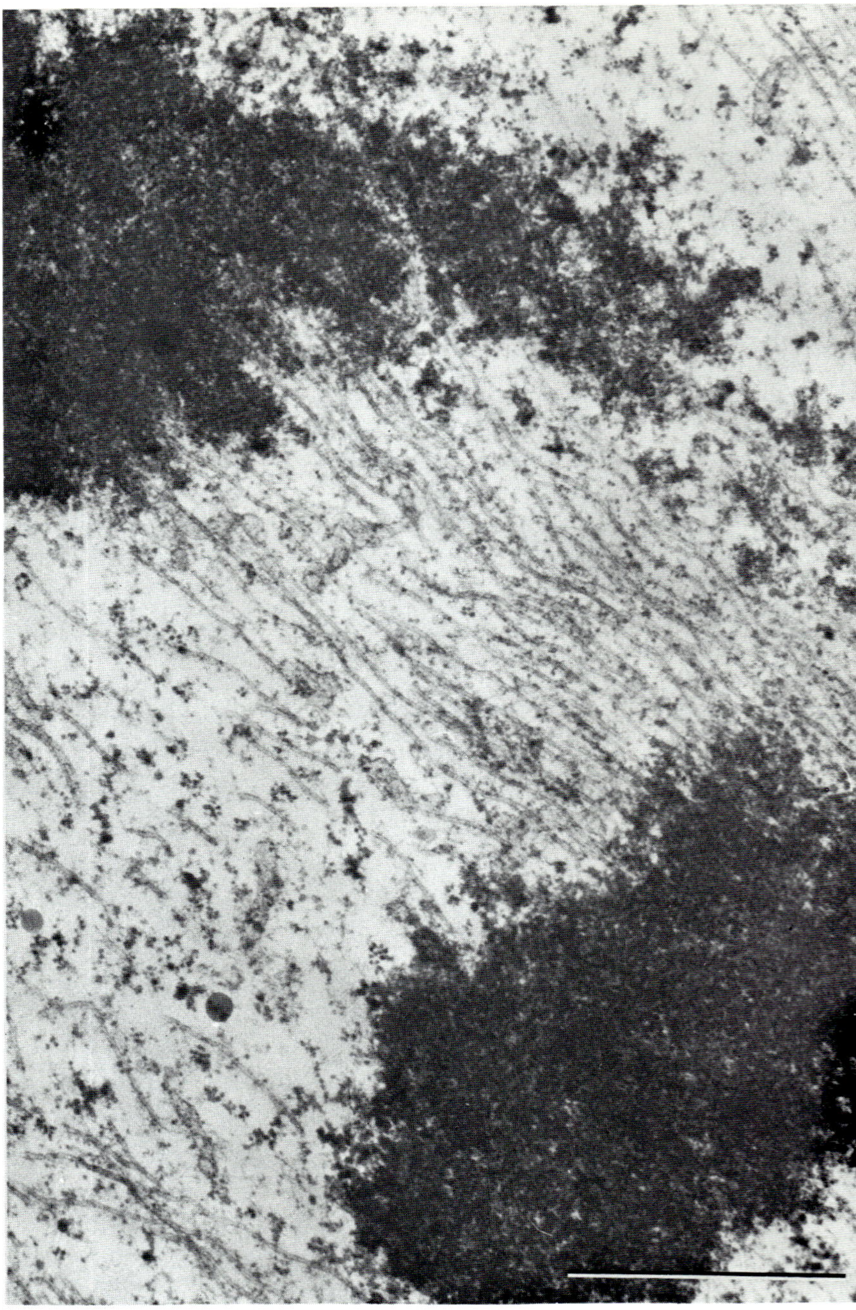

regions of the chromosomes (Inoué and Bajer, 1961; Inoué, 1964; Forer, 1966; Inoué and Sato, 1967). Similarly, most dividing cells fixed and sectioned for electron microscopy show microtubules non-uniformly distributed in the spindle; many of them are grouped in bundles associated with the kinetochores. These observations can be explained in one of two ways: either (1) the kinetochores play an active role in the assembly and orientation of kinetochore microtubules, as assumed for example in the models of Inoué (1964) and Dietz (1969); or (2) kinetochores merely act somehow to stabilize some of the microtubules or other anisotropic materials that are first formed and oriented independently; for example, at the poles of the spindle or somewhere in the interpolar region.

There is considerable evidence that kinetochores are not merely points of attachment for already formed and oriented microtubules. Kinetochores appear to play some more active role in these processes. Microscopic images have occasionally been seen that can be interpreted as showing spindle fibers running directly from one kinetochore to another. In a light microscope study of *Tradescantia* microsporocytes recovering from hydrostatic pressure treatment, Pease (1946) found three cases of what looked like "fused traction fibers" connecting the kinetochore region of one bivalent with that of another. In prometaphase of the second meiotic division in *Urechis* eggs, Luykx (1965b) found a few microtubules that appeared to connect the two sister kinetochores of one chromosome. Images suggesting microtubular connections between nonsister kinetochores have been seen by Bajer at prometaphase in *Haemanthus* endosperm cells (Fig. 6). At the very least, this kind of observation shows that kinetochores can influence the orientation of spindle fibers; it could mean that kinetochores are sites of formation of some of the microtubules.

Single chromosomes that are separated from the spindle because of spontaneous or induced division abnormality may nevertheless become associated with a few oriented spindle fibers and form small independent chromosomal spindles (Juel, 1897; Darlington and Thomas, 1937; Koopmans, 1958; Snyder, 1961; Ting, 1966; Dietz, 1963, 1969). Equatorial orientation and anaphase disjunction of such chromosomes

---

FIG. 6. Microtubules apparently connecting the kinetochores of two different chromosomes in endosperm of *Haemanthus katherinae*. (The marker represents 1 μ.) (Unpublished micrograph courtesy of A. Bajer.)

appear to be normal. Only rarely do regions of oriented spindle fibers occur separate from chromosomes (e.g., cytasters; see also Inoué, 1964, Figs. 59–62). Orientation of spindle material around individual chromosomes occurs as part of normal spindle development in some coccid insects (Hughes-Schrader, 1948b). For example, in the first meiotic division in oocytes of *Acroschismus wheeleri* (Hughes-Schrader, 1924) and in spermatocytes of *Nautococcus schraderae* (Hughes-Schrader, 1942), the first spindle fibers to appear are associated with individual chromosomes, oriented perpendicular to the opposite faces of each pair of sister chromatids in spite of the random orientation of the chromosomes with respect to one another. Only later do the individual chromosomal spindles line up in parallel and in register to form the equatorial arrangement characteristic of metaphase. Similar observations have been made in early division stages of tipulid spermatocytes by means of polarizing microscopy (Dietz, 1963, 1969). These findings suggest a direct role of the kinetochores in forming, or at least orienting, spindle microtubules.

Observation of spindle fibers elongating from the kinetochore would constitute fairly convincing evidence that the kinetochore plays a role in the assembly and orientation of spindle material. Pease (1946) interpreted his observations on *Tradescantia* microsporocytes recovering from pressure treatment this way, commenting that in some cells the traction fibers "would be thin and short, in others longer and more massive. Thus the traction fibers appeared to 'grow' outward directly away from the kinetochore region, and full thickness was not reached until they practically reached the poles." Roth and his co-workers have suggested that in amebas, during the normal course of spindle formation and in spindle re-formation after brief cold treatment, spindle microtubules gradually extend outward from the chromosomes toward the poles (Roth and Daniels, 1962; Daniels and Roth, 1964; Roth, 1967). Similarly, in *Haemanthus* endosperm cells forming spindle fibers after chloral hydrate treatment, early stages appear to show only short microtubules attached to kinetochores, while in later stages the kinetochore microtubules are longer (Molè-Bajer, 1969).

It might be argued that all of these observations, being made on fixed and sectioned cells, could as easily be the result of cell-to-cell variation or variation in the plane of sectioning relative to the spindle axis. Nevertheless, the conclusion that the kinetochore spindle fibers are assembled at the kinetochore itself also finds some support in studies on living cells irradiated with UV microbeam and observed by polarizing microscopy. When a small region between the chromosomes and the polar region in a *Haemanthus* endosperm cell is briefly irradiated, the birefringence in the irradiated region and distal to it (i.e., on the poleward side) disappears, while the birefringence next to the kinetochore is unaffected (Inoué, 1964). Over the next 10 minutes the birefringence gradually extends from the kinetochore back into the irradiated region. If the kinetochore itself is included in the irradiated region, then the whole length of the birefringent fiber disappears and "does not reappear for a long time." Similarly, in crane fly spermatocytes at metaphase I, after brief irradiation of a short region of birefringent fiber between chromosome and pole, the birefringence appears to extend from the chromosome back into the area just irradiated as the affected region (with reduced birefringence) shifts poleward (Forer, 1965, 1966). A reasonable interpretation of these observations is that throughout prometaphase and metaphase spindle fiber material is being continuously assembled and oriented at the kinetochore. The alternative—that under these experimental conditions new material is added on to the ends of already existing structure, as may be the case in flagellar regeneration (Rosenbaum and Child, 1967; Rosenbaum *et al.*, 1969)—cannot be excluded, however. The effects of irradiating only the kinetochore region on the disappearance and reappearance of spindle fiber birefringence should be studied.

While not entirely conclusive, all these observations taken together make a fairly strong case for the contention that spindle fibers (bundles of spindle microtubules) are assembled and oriented at the kinetochore during prometaphase and metaphase.

Brinkley and his co-workers (Brinkley and Stubblefield, 1966; Brinkley *et al.*, 1967) observed that in Chinese hamster cells treated with colcemid kinetochores facing the centrioles were associated with spindle microtubules, whereas the sister kinetochores (facing away from the centrioles) were invariably free of microtubules. It would be desirable to have more extensive observations to show that such images are not due to variations in the plane of sectioning relative to the orientation of the microtubules, or to asynchrony in the attachment of microtubules to the two sister kinetochores. Another seemingly contradictory observation is that in prometaphase kinetochore microtubules are not always perpendicular to the outer face of the kinetochore, as might be expected if the microtubules are organized by the kinetochore; rather, kinetochore microtubules are often approximately parallel to the interpolar axis regardless of the orientation of the kinetochore itself (e.g., Luykx, 1965b; Bajer and Molè-Bajer, 1969). The significance of these findings is not clear, but they do raise the possibility that the microtubules attached to kinetochores represent a mixture of microtubules formed outside the nucleus in prophase and microtubules organized by kinetochores in prometaphase (see also Specht, 1961).

None of these observations is really decisive one way or the other. In any case, they can be accommodated to the hypothesis that kinetochores do organize spindle fibers in many cell types, as will be discussed below in connection with the problem of chromosome orientation during prometaphase (Chapter IV, Section C,4).

In short, it is clear that the kinetochore plays some role in the establishment of kinetochore fibers (i.e., spindle fibers attached to the kinetochore), even if it is not clear what that role is. The kinetochore is also important for normal chromosome behavior during the time from early prometaphase through late anaphase. There is no convincing evidence, however, that the kinetochore plays any essential role in chromosome behavior at any other time in the life cycle of the cell.

C. KINETOCHORE STRUCTURE

1. The Primary Constriction

The kinetochore region is usually identified in fixed mitotic cells as a short nonstaining "gap," a short, thin thread connecting the two arms, or a constriction encircling each chromatid. This appearance of the site where spindle fibers meet the metaphase chromosome, the "primary constriction," may be exaggerated by certain kinds of fixation (see Prokofieva, 1935) but is not entirely an artifact, for it has also been clearly observed in living cells by phase contrast, polarizing, and differential interference microscopy (Boss, 1954; Inoué and Bajer, 1961; Bajer, 1966a). The constriction at the kinetochore region is also visible in electron micrographs of whole mounts of chromosomes spread on a water surface (e.g., DuPraw, 1966, 1968; Wolfe and Martin, 1968; Abuelo and Moore, 1969). It can be seen in prophase and anaphase chromosomes but is often less conspicuous at these stages, in the former because the relatively uncondensed arms of the prophase chromosome do not differ much in diameter from the kinetochore region itself, and in the latter possibly because of poleward extension of the kinetochore region and apposition of the proximal regions of the trailing chromosome arms.

As noted in Chapter III, Section B,4, the primary constriction at metaphase often appears to encircle each chromatid independently and symmetrically, so that in the kinetochore region sister chromatids appear not to be in direct contact.

2. DNA in the Kinetochore Region

As far as can be determined from microscopic observation, the kinetochore region contains no chemical or structural discontinuities in the elementary fiber (or fibers) that make up the chromosome. In other words, the basic structural element of the chromosome, a thread consisting mainly of DNA and protein, appears to run without interruption through the kinetochore region. Careful light microscope observation of large, Feulgen-stained chromosomes reveals that the nonstaining gap representing the kinetochore region is in fact traversed by up to four very thin, Feulgen-positive threads (e.g., Lima-de-Faria, 1956; Trosko and Wolff, 1965). Whole chromosomes from cells spread on a water surface and viewed in the electron microscope reveal in the kinetochore region a number of 200–300 A fibers indistinguishable from, and continuous with, those that make up the arms of the chromosome (DuPraw, 1966, 1968; Wolfe and Martin, 1968; Abuelo and Moore, 1969).

If the DNA of the chromosome extends through the kinetochore region, treatment with deoxyribonuclease should cause breaks in that region. However, deoxyribonuclease usually disrupts the structure of the chromosome to such an extent that the kinetochore is no longer distinguishable. [In some cases the general region of the kinetochore has proved to be more resistant than other regions to the action of deoxyribonuclease in causing breaks or removing Feulgen-positive material (see Yamasaki, 1961; MacGregor and Callan, 1962; Luykx, 1965a; Abuelo and Moore, 1969). This is attributable to tighter packing of chromatin or to the combination of DNA with other materials in these regions. In any case, these observations can be referred to the regions of the chromosome immediately adjacent to the kinetochore, the proximal heterochromatin characteristic of the chromosomes of many organisms, and may have no bearing on the question of the continuity of DNA through the kinetochore region itself.]

Digestion of chromosomes with proteolytic enzymes (MacGregor and Callan, 1962; Trosko and Wolff, 1965; Abuelo and Moore, 1969), while causing a loosening and separation of the chromatin fibers, does not produce breaks in the kinetochore region. Treatment of meiotic cells with inhibitors of protein synthesis may prevent chromosome pairing and induce abnormalities in chromosome condensation and segregation, but does not result in chromosome breakage in the kinetochore region or anywhere else (Kemp, 1964; Parchman and Stern, 1969).

A large proportion of the chromosome breaks induced by hydroxylamine in Chinese hamster cells in culture are breaks in the kinetochore region (Somers and Hsu, 1962). Since hydroxylamine reacts with cytosine in DNA, this observation could be taken as support for the idea that DNA is continuous through the kinetochore region. It is not certain, however, that DNA is the site of action of hydroxylamine in producing chromosome breaks; Borenfreund *et al.* (1964) have suggested that hydroxylamine may act to produce breaks by cleaving amino acid ester bonds in peptides that link polynucleotide chains end to end (see also Bendich and Rosenkranz, 1963). It is also not certain that a break produced by hydroxylamine in the kinetochore region is actually a break within the site of attachment of spindle fibers. If so, then one would expect both fragments to be capable of normal movement on the spindle in subsequent divisions, as in cases of kinetochore misdivision (see Chapter III, Section C, 4,*b*). Large numbers of breaks producing acentric fragments (and consequent cell mortality) and the use of colchicine in the first divisions after hydroxylamine treatment would preclude such observations, but the question might be answered by determining the karyotypes of clones from progenitor cells treated with low levels of hydroxylamine.

In short, studies of the effects of inhibitors and enzymes on the kinetochore region are inconclusive, but at least are not inconsistent with the supposition that DNA is continuous through the kinetochore region; the best support for this conclusion comes from morphological observations on whole chromosomes.

3. The Kinetochore as a Gene Product

In some light microscope preparations, particularly those fixed in osmium tetroxide and stained with iron hemotoxylin, the kinetochore region of each chromatid can be seen to contain a small differentially staining granule, termed the "spindle spherule" or "*Leitkörperchen.*" Spindle fibers (bundles of microtubules) attach to this granule rather than to the whole region of the primary constriction (see Metzner, 1894; Nawaschin, 1927; Schrader, 1936, 1939; Trankowsky, 1939; Benoit and Kehl, 1939; Iwata, 1940). It is this granule that corresponds to the kinetochore proper visualized in electron micrographs of sections as electron-dense or osmiophilic material at sites where spindle microtubules meet the chromosome (see Luykx, 1965a). Changes in the amount and extent of this material as division progresses (Schrader, 1936, 1939; Luykx, 1965a; Jokelainen, 1965 and unpublished; Pickett-Heaps and Fowke, 1969), as well as its differential stainability, suggest that at least some of this material represents a product of genes embodied in the DNA of the kinetochore region. This DNA may therefore be viewed as a "kinetochore organizer," similar to the nucleolar organizing region of the chromosome in a number of ways. It is responsible for the synthesis or assembly of an essential organelle that remains associated with a specific chromosomal site, is often associated with blocks of heterochromatin, remains relatively uncoiled during mitosis, and probably contains a large number of identical genes.

4. Evidence for Many Gene Copies in the Kinetochore Region

The idea that the kinetochore region contains more than one copy of the gene responsible for the attachment of spindle microtubules finds some support in direct microscopic observation and in observations on the consequences of transverse breakage through the kinetochore region.

a. Structure

Within the kinetochore region of each chromatid of large chromosomes, two or occasionally even three Feulgen-staining "centromeric chromomeres" can sometimes be observed (Iwata, 1940; Tjio and Levan, 1950a,b; Lima-de-Faria, 1958a,b; MacGregor and Callan, 1962). Lima-de-Faria (1956) has described the structure of the kinetochore region in pachytene microsporocytes of a number of plants as a "tandem reversed repeat." The relation of these chromomeres to the kinetochore itself is not clear, but nevertheless their arrangement does suggest a compound structure, particularly when kinetochores located interstitially are compared with terminal kinetochores. For example, the single terminal chromomeres of the telocentric chromosomes of *Tradescantia micrantha* appear to be doubled in the interstitial kinetochore regions of *Allium cepa* chromosomes (Jones and Colden, 1968). John and Hewitt (1966) have made similar observations, comparing telocentric and metacentric chromosomes in certain species of short-horned grasshoppers.

b. Transverse Breakage

A transverse break in the kinetochore region can produce two chromosome fragments both of which are capable of normal behavior during cell division. In two individual corn plants, among 466 progeny of a cross involving X-irradiated pollen, McClintock (1938) found instead of a normal paternal chromosome V two chromosomes, a small ring-shaped chromosome and a chromosome V deficient for a segment of its short arm. Both possessed functional kinetochores, as indicated by their normal behavior during mitosis in the two original plants and their offspring.

In the grasshopper *Stauroderus scalaris*, John and Hewitt (1968) found in one individual a group of primary spermatocytes in which two whole-arm translocations were present, accompanied by two minute fragments. These minute fragments apparently had functional kinetochores because they were regularly present in the cells with the altered chromosomes. The metaphase configurations of the whole-arm translocation products indicated that they too had functional kinetochores. The presence of these altered chromosomes with the centric fragments increased the normal number of kinetochores in these cells by two. This can only mean that in the original cell that produced this group of spermatocytes one of the chromosomes involved in each translocation must have fractured through the kinetochore region to give two altered chromosomes, each with a functional kinetochore.

Similarly, in one individual in a population of the grasshopper *Myrmelotettix maculatus*, Southern (1969) found that one member of a pair of homologous metacentric chromosomes had been replaced by two telocentric chromosomes. The pairing behavior of these chromosomes in meiotic prophase leaves no doubt that the telocentrics were indeed homologous with the two arms of the metacentric. Since the telocentrics behaved normally during the meiotic divisions and were present in all the spermatocytes examined, their kinetochores must have been functioning normally.

The production of two functional kinetochores from one is also suggested by the phenomenon of kinetochore "misdivision" (Upcott, 1937; Koller, 1938; Darlington, 1939; see review by John and Lewis, 1965, pp. 52–62). Unpaired chromosomes in the first meiotic division often remain at the equator of the spindle during most of anaphase. In late anaphase or early telophase, they may undergo breakage and abnormal division such that two *sister* chromosome arms move toward one pole while the other two sister arms move toward the opposite pole. Similar events may also occur occasionally in abnormal mitotic divisions. In meiotic misdivision both of the "new" chromosomes thus formed may behave normally in the second meiotic division, with sister arms separating and moving to opposite poles during anaphase (Darlington, 1939). A single chromatid from normally dividing univalents may also undergo misdivision, the two arms moving in opposite directions. Often a thin strand of chromatin connects the separating chromosome arms as they move to opposite poles, as if a portion of the kinetochore region itself were being stretched between two opposed segments of the same kinetochore.

Although it is not known in any instance whether or not the four independent products of a meiotic misdivision can *all* behave normally in postmeiotic mitoses, Sears' (1952) cytogenetic studies on wheat showed that at least some of them are transmitted to the next generation and can behave normally during mitosis in the growth and development of the progeny.

c. Estimate of the Amount of DNA

If it is assumed that the chromatin fibers in the kinetochore region have the same general composition as those elsewhere in the chromosome, a rough estimate can be made of the amount of DNA in the kinetochore region of human chromosomes. DuPraw (1968) has estimated that there are five to seven 250-A chromatin fibers running through the constricted kinetochore region and that each chromatin fiber contains about 1.8×10^{-16} gm of DNA per micron. If the length of each chromatin fiber in the kinetochore region is about 0.5 μ [see DuPraw's Fig. 18-7 (A)], then the kinetochore region would contain about 5×10^{-16} gm of DNA, or 500,000 nucleotide pairs. All kinetochore regions of the human diploid set of chromosomes together would contain about 0.5% of the total DNA. While several gene products could be involved in kinetochore structure and function—in differential condensation or coiling, in the organization of or attachment to spindle microtubules—the relatively large amount of DNA present (enough to code for approximately 1000 different proteins), taken together with the conclusion that a break through the kinetochore region can yield two fragments each with a functional kinetochore, suggests a high degree of gene duplication in the kinetochore region.

5. Are All Kinetochores of a Chromosome Set Alike?

There are a few observations that suggest functional differences between kinetochores within a set of chromosomes. Barber (1941) observed in pollen grains of *Uvularia* 3 metaphases, out of more than 200, in which one of the seven chromosomes (morphologically distinguishable) showed a delay or failure in movement onto the metaphase plate, and in each case it was the E chromosome. In a careful study of the chromosome elimination that normally occurs in the third and fourth cleavages of embryos of the gnat *Miastor,* Nicklas (1959) found that all chromosomes behave normally up through early anaphase, with all kinetochores orienting normally and beginning normal anaphase movement toward the poles. The chromosomes to be eliminated, however, fail to complete their poleward movement. Failure of complete separation at the distal ends of the chromosome arms cannot explain the observations, for in many cases the chromatids appear to have separated completely yet fail to reach the poles. Hughes-Schrader (1950) observed a somewhat similar lagging of the sex chromosomes in midanaphase of the spermatogonial divisions in the mantid *Stagmatoptera septentrionalis.* The differential behavior of maternal and paternal chromosomes in the monopolar division of *Sciara* spermatocytes (Chapter IV, Section E,2) and the differential effect of the two members of a heteromorphic autosomal bivalent on sex chromosome orientation in *Gryllotalpa* spermatocytes (Chapter IV, Section E,3) also suggest regular functional differences between kinetochores in these cells.

Possible functional differences between kinetochores of different chromosomes might be assessed by observing the meiotic behavior of certain interchange heterozygotes (e.g., Lewis and John, 1963), in which nonhomologous kinetochores linked by chiasmata may be oriented toward opposite poles, but there do not appear to have been any sufficiently detailed observations of chromosome behavior under these circumstances to allow any conclusions.

There is no concrete evidence that where differential behavior of particular chromosomes within a set exists it is caused by genetic differences in the kinetochores themselves. It could as well be the result of differences in other regions of the chromosome, differences that only indirectly affect kinetochore function. Factors similar to those determining differential condensation or "facultative heterochromatization" (see Brown, 1966) may be involved.

Where this question has been investigated in any detail, differential chromosome behavior during division has indeed been shown to be due to chromosome segments distinct from the kinetochore region. Preferential segregation of one homolog of chromosome 10 in megasporogenesis of maize is attributable not to special properties of its kinetochore but to a heterochromatic segment or "knob" at the distal end of the long arm (Rhoades, 1942, 1952). In oocytes of *D. melanogaster* a dicentric X chromosome (produced by crossing-over in inversion heterozygotes) ordinarily forms a bridge, in anaphase of the first meiotic division, that prevents either of the opposed kinetochore regions from carrying the chromosome into the egg nucleus. Certain combinations of X chromosomes, however, yield asymmetrical dicentrics that may reach the egg nucleus intact, suggesting that the kinetochores of some X chromosomes can overcome the opposing force of the kinetochores of other X chromosomes (Novitski, 1952, 1955; see also Ptashne, 1960). Analysis of the effects of a variety of chromosome rearrangements has shown that such differences in the kinetic behavior of the X chromosome do not result from differences between kinetochores but from differences in the adjacent chromosome segments (Lindsley and Novitski, 1958). Similarly, Crouse (1960) has shown that nondisjunction and "precocious" movement of the sex chromosome in secondary spermatocytes of *Sciara* is caused by a short heterochromatic segment adjacent to the kinetochore, not to the kinetochore itself. When this heterochromatic segment is translocated to the distal end of a second chromosome, then it is this second chromosome that shows the characteristic nondisjunction and precocious behavior. All these cases are most simply explained by assuming the development of "neocentric" activity by chromosome segments other than those containing the kinetochores, as the cytological observations of Rhoades (1952) have shown for maize.

6. Electron Microscopy

The most detailed descriptions of kinetochores are from sections of fixed cells, particularly from the observations of Jokelainen (1967)

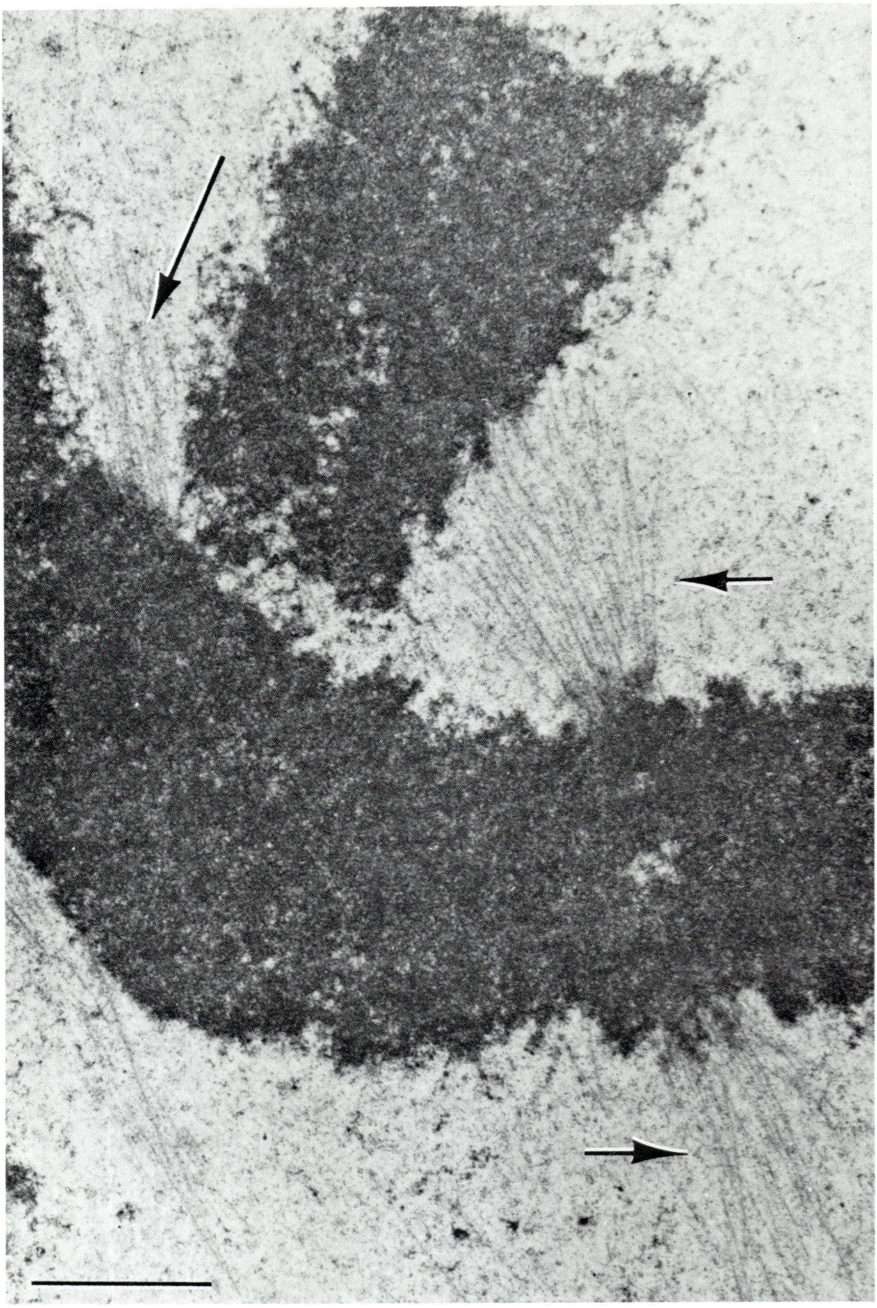

and Brinkley and Stubblefield (1966) on mammalian cells, Bajer (1968a,b) on endosperm cells of the liliaceous plant *Haemanthus*, and H. J. Wilson (1968) on pollen mother cells of *Tradescantia*. The chromosomal structures to be described are not simply a response of chromosome material to association with spindle microtubules, for they are seen also in prophase, before the nuclear envelope breaks down and microtubules become associated with chromosomes, and in cells treated with mitotic inhibitors, in which organized spindle microtubules are absent. Furthermore, the typical details of kinetochore structure are not seen at other sites on the chromosome to which spindle microtubules may become temporarily attached (neocentric fibers; see Fig. 7, from Bajer, 1968b).

a. Animal Cells

In Jokelainen's (1967) study of metaphase in three cell types in the rat embryo, the kinetochore appears as a complex structure in the primary constriction at the surface of the chromosome. It consists of an outer finely filamentous or amorphous dense layer (facing the pole) 30–45 mμ thick, a middle layer 15–35 mμ thick, of low density, and a dense inner layer, 15–30 mμ thick, made up of relatively coarse granules or fibers continuous with those of the main body of the chromosome (Figs. 8 and 9). Sometimes an amorphous material of low density is seen on the outer face of the outer layer (Fig. 8). In section these layers appear as bands 200–250 mμ long at the surface of the chromosome. That these bands represent layers of a disk-shaped structure is shown by serial sectioning of individual kinetochores, by the fact that the bands have similar dimensions whether the chromosome is sectioned in the spindle axis longitudinally or transversely, and most directly by face views, in which the kinetochore is a circular plate 200–250 mμ in diameter (see Fig. 10). These observations also show that the dense layers do not encircle the metaphase chromosome. Sister kinetochores are restricted to small areas on opposite sides of the chromosome. Each

---

FIG. 7. A neocentric fiber (bundle of microtubules indicated by long arrow) attached to a chromosome in an endosperm cell of *Haemanthus katherinae*. The two sister kinetochore fibers are indicated by the short arrows. Note also the small hole separating the sister chromatids in the kinetochore region. (The marker represents 1 μ.) (From Bajer, 1968b. Copyright 1968 by Springer-Verlag, Berlin-Heidelberg-New York.)

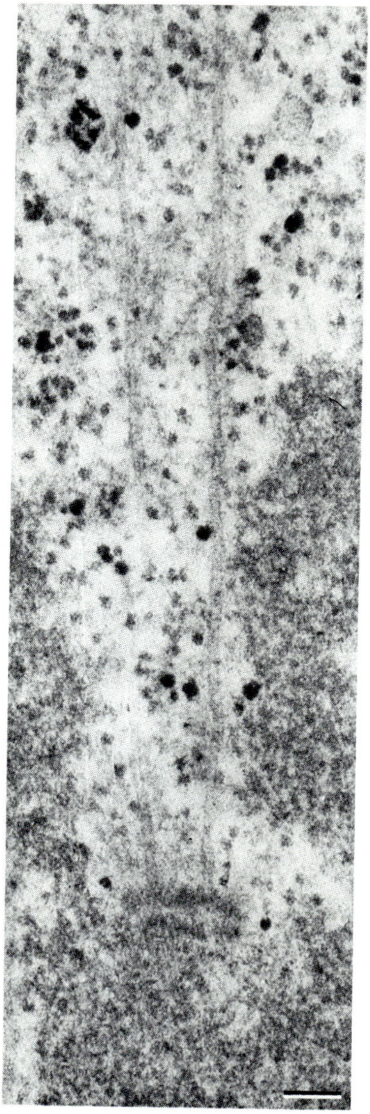

FIG. 8. Kinetochore in a rat embryo cell at metaphase, showing two osmiophilic or electron-dense layers separated by an intermediate lighter layer. Note also the faint amorphous corona of material surrounding the microtubules on the outer face of the kinetochore. (The marker represents 0.1 μ.) (From Jokelainen, 1967. Copyright 1967 by Academic Press, Inc.)

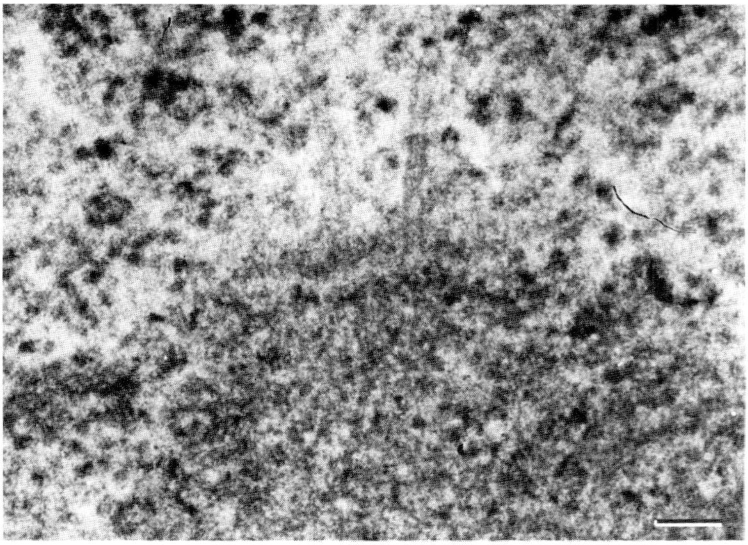

FIG. 9. Kinetochore in a rat embryo cell at metaphase. One microtubule penetrates the outer dense layer; a few thin filaments appear to extend from the inner dense layer downward into the body of the chromosome. (The marker represents 0.1 μ.) (From Jokelainen, 1967. Copyright 1967 by Academic Press, Inc.)

kinetochore appears to be more or less surrounded by chromosome material so that only its outer face is directly exposed to materials in the spindle.

There is little discernible differentiation in the region underlying the kinetochores, although sometimes the chromatin fibers in this region look more compacted or more densely stained than elsewhere in the chromosome, and occasionally there are suggestions of thin filaments from the kinetochore penetrating the underlying regions (Fig. 9).

Microtubules from the poles traverse the outer two layers and appear to terminate at the inner dense layer (Fig. 9). It is conceivable that some microtubules continue as the thin filaments past the inner dense layer and hence actually penetrate more deeply into the body of the kinetochore region.

A number of the features of kinetochore structure in cells of the rat embryo just described have also been observed in other cells. The

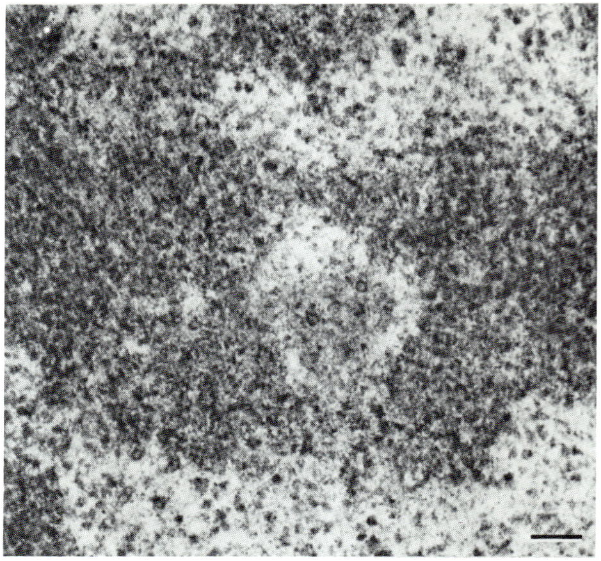

FIG. 10. Face view of a metaphase kinetochore in a rat embryo cell. Note the two or three cross-sectioned microtubules in the kinetochore. (The marker represents 0.1 μ.) (From Jokelainen, 1967. Copyright 1967 by Academic Press, Inc.)

kinetochore has clearly been shown to be restricted to a small area on each chromatid also in young sea urchin embryos (Harris and Mazia, 1962), in eggs of the echiuroid worm *Urechis caupo* (Luykx, 1965b), and in Chinese hamster cells in culture (Brinkley and Stubblefield, 1966). In addition to their presence in the rat embryonic cells described by Jokelainen (1967), distinct layers of the kinetochore occur in Chinese hamster cells in culture (Brinkley and Stubblefield, 1966), mouse L-strain fibroblasts (Krishan and Buck, 1965), HeLa cells (Robbins and Gonatas, 1964), pigeon spermatocytes (Nebel and Coulon, 1962), chick embryo cells (Allenspach and Roth, 1967), eggs of *U. caupo* (Luykx, 1965a), cells of the multicellular green alga *Oedogonium* (Pickett-Heaps and Fowke, 1969), and in avian malarial parasites (Aikawa and Beaudoin, 1968). The layered structure of the kinetochore probably occurs also in the rabbit egg (Zamboni and Mastroianni, 1966), in the cockroach *Periplaneta* (Roth *et al.*, 1966), and in the algae *Nitella* (Turner, 1968) and *Chara* (Pickett-Heaps, 1967). The inner dense layer is not always visible as a structure distinct from the underlying chromatin, possibly because of the intense general staining of the chromosomal material by uranyl actate, which is used routinely in many laboratories.

The inner dense layer is most readily interpreted as the kinetochore organizer containing the kinetochore genes. It has a rough granular or fibrous texture similar to, and is continuous with, the adjacent chromatin material. This interpretation is consistent with the observations of Wettstein and Sotelo (1965) on spermatocytes of the cricket *Gryllus argentinus*, in which the kinetochore appears to be made up of the same elementary fibrils that comprise other regions of the chromosome, the main difference being the tighter packing and the possible association of fibrils in the kinetochore region with amorphous dense material not present in other parts of the chromosome. The light middle layer, outer dense layer, and associated material may reflect kinetochore-gene activity, or the association of kinetochore-gene products with spindle microtubules. In support of the idea that the outer dense layer consists of material different from that composing the underlying region of the chromosome, it may be noted that (1) the outer dense layer has a fine filamentous or amorphous structure not as coarsely granular as the inner dense layer; (2) the electron density of the outer layer stays about the same under different staining conditions, while that of the inner layer and other regions of the chromosome does not (Jokelainen, 1967); (3) electron-dense material remains associated with the chromosome at the kinetochore region even with enzyme treatments or fixation conditions that remove substantial amounts of chromosome material (Harris and Mazia, 1962; Luykx, 1965a); and (4) the material of the outer layer appears to change in amount or degree of condensation during the progress of some cells from prophase to metaphase. The kinetochore is present as a differentiated region on the surface of the chromosome already in prophase, before the nuclear envelope breaks down, and before spindle microtubules are attached to the chromosome (Schrader, 1936, 1939; Brinkley and Stubblefield, 1966; Jokelainen, 1967). As the cell progresses toward metaphase, there is a decrease in the surface area occupied by the kinetochore and an increase in its density (Schrader, 1936, 1939; Luykx, 1965a; Jokelainen, 1965 and unpublished observations). Similar changes have also been described for the kinetochore in the hypermastigote flagellate *Barbulanympha* (Hollande and Valentin, 1968a).

Brinkley and Stubblefield (1966) have described the fine structure of the kinetochore in Chinese hamster cells dividing synchronously after colcemid treatment. Their observations are similar to those described above, except that the outer dense layer (20–30 mμ thick) appears to be composed of three sublayers, two dense sublayers separated by a less dense sublayer, each 5–8 mμ thick. Furthermore, the outer dense layer in some sections extends along the surface of the chromosome for a distance of 500 mμ, twice the size of the metaphase kinetochores in the dividing rat embryo cells described by Jokelainen. The larger size of the outer dense layer and its separation into three sublayers may be a characteristic of kinetochores in cells treated with mitotic inhibitors, since these features have also been seen in HeLa cells treated with vincristine (George *et al.*, 1965) and in mouse L cells treated with vinblastine (Krishan, 1968). There is some evidence (Jokelainen, 1968) that colchicine retards the condensation of dense kinetochore material that normally occurs in the transition from prophase to metaphase.

Brinkley and Stubblefield (1966) offer an interpretation of kinetochore structure substantially different from that presented above. First, they interpret the outer dense band not as a section of a disk but as a filament, or a pair of intertwined filaments, that arises from the surface of the chromatid at one point, runs along the chromatid for a short distance, and inserts at a point farther along on the same chromatid. They view the middle layer and the "corona" on the kinetochore's outer face as a system of fine microfibrils extending laterally from the kinetochore filament and liken the structure of the whole kinetochore to the arrangement of lateral loops along a main axis seen in lampbrush chromosomes. This interpretation seems less likely than the one outlined in the preceding paragraphs, because the kinetochore presents the same structure in chromosomes sectioned (parallel to the spindle axis) longitudinally or transversely, and because no filamentlike structure has been clearly seen in face views of the kinetochore (Fig. 10). Furthermore, that the outer layer does not respond to staining and extraction procedures in the same way as do fibers in other regions of the chromosome would suggest that this material is not continuous with the fibers of the chromosome itself.

Because the size of many of these structures is smaller than the thickness of sections used for electron microscopy, it is not likely that sectioned material will give much more information on the details of kinetochore structure and its relation to the chromosome. The technique of spreading whole cells on a water surface allows microtubules to be preserved intact and still connected with the chromosome (Barni-

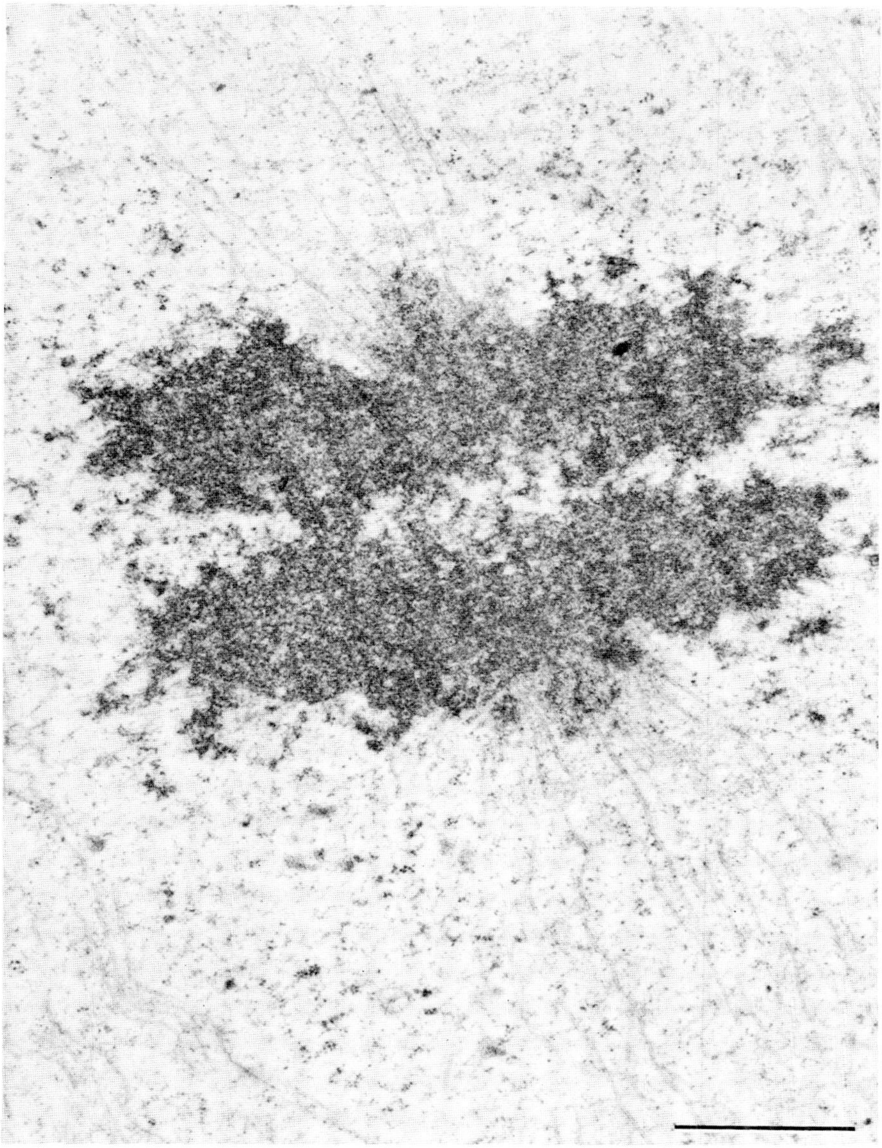

FIG. 11. Two sister kinetochores in an endosperm cell of *Haemanthus katherinae*. (The marker represents 1 μ.) (Unpublished micrograph courtesy of A. Bajer.)

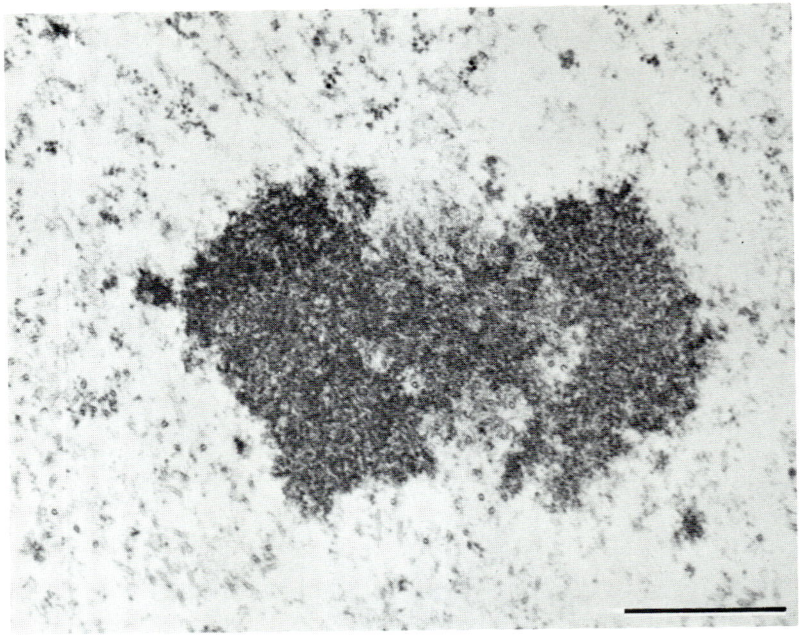

FIG. 12. Face view of a kinetochore in an endosperm cell of *Haemanthus katherinae*. Several cross-sectioned microtubules associated with the kinetochore can be seen. (The marker represents 0.5 μ.) (From Bajer, 1968a).

cot, 1966); perhaps a refinement of this method will reveal additional details of the structure of the kinetochore and its association with spindle microtubules.

b. Plant Cells

The large chromosomes in the cells of higher plants that have been investigated also possess kinetochores that are restricted to a relatively small area on the surface of each chromatid, as judged from longitudinal and transverse sections of chromosomes and from sections perpendicular to the spindle axis showing kinetochores in face view (Bajer, 1968a,b; H. J. Wilson, 1968; see Figs. 11, 12). Like the kinetochores of animal cells, they contain material intimately associated with the ends of spindle microtubules, material that stains differently from the rest of the chromosome, and they are recognizable in prophase, before the nuclear envelope breaks down (Bajer and Molè-Bajer, 1969).

There are, however, a number of differences between the structure of kinetochores in plant cells and the disklike structure of animal cell kinetochores. The following description is based on the observations of Bajer (1968a,b) on *Haemanthus* endosperm cells and those of H. J. Wilson (1968) on *Tradescantia* pollen mother cells in the second meiotic division, but it probably applies as well to pollen mother cells of *Lilium* (Dietrich, 1966, 1968), leaves of *Nicotiana* (Cronshaw and Esau, 1968), and sporogenous tissue of *Psilotum* (Allen and Bowen, 1966) and *Equisetum* (Manton, 1964a). The kinetochore of plant cells in mitotic metaphase or anaphase is best described as a "ball-and-cup" structure. The chromosome material in the kinetochore region is raised into a

FIG. 13. Details of the association between microtubules and a chromosome in a dividing endosperm cell of *Haemanthus katherinae*. Long arrows, bundles of kinetochore microtubules; short arrows, penetrating microtubules. (The marker-represents 0.5 μ.) (From Bajer, 1968b. Copyright 1968 by Springer-Verlag, Berlin-Heidelberg-New York.)

circular rim that surrounds a central depression, thus forming a shallow cup 1.2–1.7 μ in diameter on the surface of the chromosome (see Figs. 7, 11). In this cup lies a ball up to 0.7–0.8 μ in diameter (two or three times the diameter of the metaphase kinetochore of mammalian cells) composed of a fine filamentous or granular material that stains less densely than the body of the chromosome. Spindle microtubules appear to penetrate the ball to various depths; a few can be traced to the bottom of the cup (Fig. 13). The ball often appears to make direct contact with the chromosome only in a small region at the bottom of the cup. Unlike the kinetochore of the mammalian chromosome, the kinetochore here appears to be exposed to the spindle not only on its outer surface but also on the sides, for transverse sections of chromosomes and face views of kinetochores (Dietrich, 1966; H. J. Wilson, 1968; Bajer, 1968a,b; see Fig. 12) indicate that at right angles to the chromosome axis the rim of the cup is incomplete.

Like the outer dense layer of the kinetochore of animal cells, the ball of the kinetochore in these plant cell chromosomes is the primary site of attachment of chromosomal microtubules, stains differently from the rest of the chromosome and, as Molè-Bajer (1969) has noted in *Haemanthus* endosperm cells, becomes more conspicuous as spindle microtubules develop in prometaphase and metaphase. It is therefore probably the functional counterpart of the outer dense layer of the kinetochore in animal cells.

It should be noted in addition that the ball-and-cup structure of the kinetochore may be a characteristic not of higher plant cells in general but rather of cells with large chromosomes, for a similar structure was described by Schrader (1936, 1939) in spermatocytes of *Amphiuma* and by Barnicot and Huxley (1965) in cultured cells of *Triturus*.

c. Barbulanympha

The kinetochores of the flagellate *Barbulanympha* (see Hollande and Valentin, 1968a) are unusual in their large size (1–3 μ in diameter during interphase, gradually condensing to a dense structure about 0.5 μ in diameter and 0.2–0.4 μ thick at division) and in their insertion into the intact nuclear envelope at division. Nevertheless, they show a number of close similarities to the kinetochores of metazoan cells: (1) they undergo a progressive condensation during the early stages of division; (2) they consist of a number of electron-dense disks or layers, the outermost one being a thick cup-shaped structure covered on its outer surface with a corona of more lightly staining material penetrated by the spindle microtubules; (3) the innermost disks or layers consist of compacted fibers that appear to be continuous with the fibers of the chromosome itself; and (4) a core of oriented fibers extends from the kinetochore into the body of the chromosome, similar to the core described by Wettstein and Sotelo (1965) in cricket spermatocytes and by Harris and Mazia (1962) in partially extracted mitotic chromosomes of sea urchin embryos.

d. Chromosomes without Well-Defined Kinetochores

In a number of microorganisms—amebas (Roth *et al.*, 1960; Roth and Daniels, 1962; Daniels and Roth, 1964; Roth, 1967), slime molds (Aldrich, 1969), and ciliates (Roth and Shigenaka, 1964; Tucker, 1967; Jenkins, 1967)—there does not appear to be any well-defined differentiated structure at sites where spindle microtubules meet the chromosome. Some of the published electron micrographs do, however, show spots of electron-dense material or light amorphous material associated with the chromosomal ends of spindle microtubules (Daniels and Roth, 1964; Roth and Shigenaka, 1964; Jenkins, 1967). Whether this inconspicuous, differentially staining material is a consistent feature of the microtubule attachment sites in these cells has yet to be determined.

Behnke and Forer have claimed that in the spermatocytes of the crane fly *Nephrotoma suturalis* the spindle microtubules run along the surface of the chromosome in grooves or channels, or pass directly through the chromosome, but do not insert on the chromosome at any differentiated site recognizable as a kinetochore. It is likely, however, that in these large cells (30–40 μ in diameter), with their small number of chromosomes ($2n = 8$), very few sections would pass through a kinetochore. Figure 5 in their paper in fact shows a bundle of microtubules inserting on a chromosome at a site morphologically similar to the kinetochore of the large chromosomes of higher plant cells. Further studies of this material at the electron microscope level would be desirable, for at the light microscope level there is little to indicate that the relation between spindle fibers and chromosomes is unconventional (see Dietz, 1969): (1) the configurations of the chromosomes in metaphase and anaphase seen in living cells by phase contrast microscopy suggest that traction fibers attach to the chromosomes at restricted sites; (2) observations of living cells with the polarizing microscope show a streak of birefringent material on the poleward side of each anaphase chromosome; (3) in spermatogonial and oogonial divisions of other members of the tipulid family, each chromosome has a small constricted region that leads the way to the poles at anaphase (see figures in Ullerich *et al.*, 1964).

Somewhat similar electron microscope observations have been made by Buck (1967) on dividing cells in *Rhodnius*, but here there are grounds for expecting that typical kinetochores will not be found. This insect belongs to the order Hemiptera, whose members characteristically possess "holokinetic" chromosomes, or chromosomes with "diffuse kinetochores." These terms reflect the observations that (1) anaphase chromosomes usually move broadside to the poles, and (2) X-ray-induced chromosome fragments behave normally throughout division. [The special problems of chromosome organization and behavior presented by hemipteran insects and other groups of organisms with holokinetic chromosomes will not be discussed here. The reader is referred to the discussion and references in a recent paper by Bauer (1967).]

The relation of microtubules to the orderly movement of chromosomes in dividing cells of *Saccharomyces* (yeast) is not clear. Interestingly, genetic analysis of a diploid strain (Mortimer and Hawthorne, 1966) has given clear evidence of at least 14 linkage groups, whose segregation at meiosis appears to be controlled by conventional meiotic kinetochore behavior; yet in the vegetative divisions of the tetraploid strain examined by Robinow and Marak (1966), where at least 56 chromosomes would be expected, the entire "fiber apparatus" of the dividing nucleus consists of not more than about *15* microtubules. These remarkable findings might be explained by assuming that the chromosomes in vegetative division aggregate to form one or a few compound chromosomes in which many of the kinetochores need not form connections with microtubules. Or they might mean that the chromosomes are distributed equally to daughter cells by some other mechanism altogether, as appears to be the case in the division of dinoflagellates (Kubai and Ris, 1969; Soyer, 1969). In these cells bundles of microtubules lie in membrane-bounded cytoplasmic channels that pass through the nucleus. The chromosomes are attached to the intact nuclear envelope on its inner surface and apparently make no direct contact with the microtubules on the other side of the nuclear envelope at any time during division. Possibly in yeast cells, as well as in dinoflagellates, the segregation of daughter chromosomes is most immediately determined by growth of the nuclear envelope in regions between attachment sites of daughter chromosomes, in a manner similar to that proposed for bacterial division (see Chapter I, Section C). Whatever the mechanism, microtubules appear to be less directly involved in chromosome segregation in these two cell types than they are in most other cells, and kinetochores cannot be cytologically defined as microtubule attachment sites.

D. KINETOCHORES IN MEIOTIC DIVISIONS

1. Kinetochore Duplication, Segregation, and Orientation

A direct corollary of the partial chiasmatype theory (see Darlington, 1965, Ch. VII; and Whitehouse, 1965, Ch. 8, for review) is that in the first meiotic division the two sister kinetochores of one homolog do not separate and move to opposite poles but move together to one pole, while the two sister kinetochores of the other homolog move together to the other pole ("reductional" segregation). The most direct evidence for this kind of segregation of kinetochores in meiosis comes from studies in which the kinetochore region of one of the homologs is "marked" by a mutant gene or by a structural change visible in the light microscope. Except for instances in which crossing-over or chiasma formation has occurred between the kinetochore region and the marker, the two chromatids bearing the marker always move to the same pole at anaphase I, the two unmarked chromatids moving to the opposite pole. Darlington's precocity theory of mitosis explained this by assuming that the chromosomes enter the first meiotic division with kinetochore regions actually unreplicated, so that the kinetochore of each member of a homologous pair can form a connection with only one pole. The answer is not quite so simple, for, as Östergren (1951) proposed in his analysis of meiotic chromosome orientation and as Luykx (1965a) showed directly, in *prometaphase I* one kinetochore region of a bivalent can in fact be connected by spindle fibers to *both* poles.

Furthermore, the evidence indicates that in meiosis as in mitosis the kinetochore regions are already duplicated when the chromosomes first move on to the spindle. The earliest meiotic stage in which duplicated kinetochores have been definitely identified is diakinesis in spermatocytes of *Amphiuma* (Schrader, 1936, 1939). By mid- or late diakinesis two adjacent "spindle spherules" or kinetochores can be

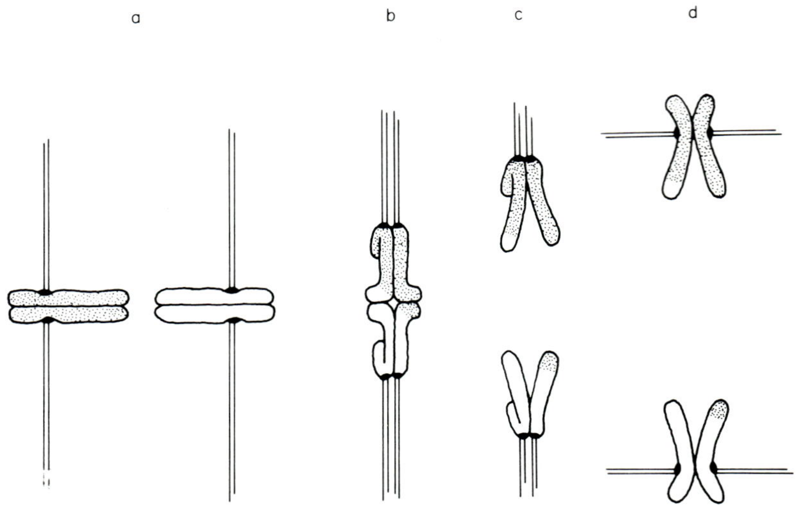

FIG. 14. The disposition and orientation of kinetochores in meiotic divisions. (a) Mitotic metaphase for comparison; (b) metaphase I, sister kinetochores adjacent; (c) anaphase I; (d) metaphase II, sister kinetochores opposite.

seen on each chromosome of the bivalent, evidently representing the two sister kinetochores organized by the underlying duplicated chromatids. The sister kinetochores remain closely associated throughout the first meiotic division, lying on the same side of the sister chromatids. By prometaphase of the second division, however, the sister kinetochores have separated and now lie on opposite sides of the chromosome (see diagram, Fig. 14). The presence of two kinetochores at each end of the bivalent at metaphase I is commonly seen in squash preparations of microsporocytes (e.g., Fig. 15). The presence of two kinetochores in each homolog at metaphase I and anaphase I and their separation before or during prometaphase II have also been documented by electron microscopy of sections of *Urechis* eggs (Luykx, 1965b). That both sister kinetochores are functional is shown by the attachment of spindle microtubules to both (Luykx, 1965a), and by the precocious separation of sister kinetochore regions that has occasionally been observed at anaphase I of male meiotic cells (e.g., Richardson, 1936; Klingstedt, 1939; Brown and Zohary, 1955; John and Hewitt, 1965, Figs. 32 and 33).

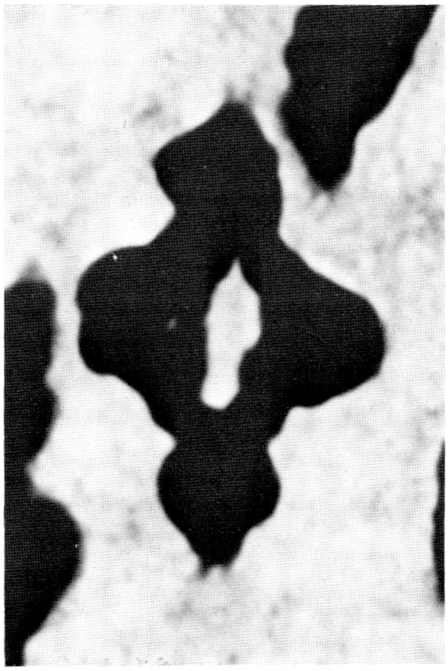

FIG. 15. Light micrograph of a bivalent from a pollen mother cell of *Lilium formosanum*. Note that there is a pair of "traction cones" (probably representing the two sister kinetochores being stretched poleward by spindle fibers) at each end of the bivalent.

A gradual separation of sister kinetochores during anaphse of the first meiotic division was described by Schrader (1939) in *Amphiuma* spermatocytes and by Lima-de-Faria (1956, 1958a) in *Tradescantia* pollen mother cells. The changes that lead to the disposition of kinetochores on opposite sides of the chromosome by the beginning of the second meiotic division may therefore actually begin during anaphase I.

2. Hypotheses Accounting for the Change in Kinetochore Orientation in Meiosis

The mechanism by which sister kinetochores are separated from adjacent positions to opposite sides of the chromosome between the first and second meiotic divisions is an important problem, for it is clearly related to the difference in the orientation of the chromatids in the two divisions. A micromanipulation experiment of Nicklas (1966) indicates that this structural difference in the kinetochore region does indeed affect chromatid orientation. Operating on a cell formed by the fusion of two grasshopper spermatocytes—one in the first meiotic division, the other in the second—he detached a chromosome from the second meiotic spindle and placed it near the first meiotic spindle. The subsequent behavior of the chromosome indicated that sister kinetochores had become attached by spindle fibers to opposite poles of the first meiotic spindle; it oriented at the equator, and at anaphase the two chromatids separated and moved to opposite poles along with the half-bivalents. Quite different is the behavior of an unpaired chromosome that enters the first meiotic spindle via a more natural route (e.g., because of a pairing failure in meiotic prophase) and which presumably has its kinetochores positioned side by side as in the normal bivalent. Such chromosomes are usually oriented toward one pole or the other throughout most or all of the first meiotic division (see Chapter III, Section D,2,*c* for further discussion of univalent behavior).

The importance of the "opposite" versus "adjacent" arrangement of sister kinetochores in determining chromosome orientation is also clearly illustrated by the difference in orientation of *mitotic* dicentric chromosomes (formed by breakage and fusion) and *meiotic* dicentric chromosomes (formed by a double crossover within a heterozygous inversion). In the mitotic dicentric the two kinetochore regions and the connecting chromosome lie in the plane of the spindle equator at metaphase, and sister kinetochores separate to opposite poles at anaphase (e.g., Bajer, 1963). In the comparable dicentric in the first meiotic division, kinetochore regions lie equal distances on either side of the equator, with the connecting chromosome segment perpendicular to the equatorial plane. One pair of sister kinetochores is oriented toward one pole, the other pair toward the opposite pole (e.g., Brandham, 1969, Fig. 5).

These observations leave little doubt that the disposition of the sister kinetochores on the chromosome surface is a major factor in determining chromosome orientation and behavior.

How is the change in kinetochore disposition between the two meiotic divisions brought about?

a. Activation of Different Sites

The kinetochore region may be arranged and controlled in such a way that in the beginning of the first meiotic division certain stretches of chromatin fibers on the same sides of the two sister chromatids are activated to form two adjacent kinetochores, while at some time between the end of the first division and the beginning of the second these are inactivated and other stretches of kinetochore-organizing chromatin fibers, those on opposite sides of the sister chromatids, become activated. There is little evidence in support of this idea, but it is consistent with the conclusions that the kinetochore region of each chromatid contains more than one kinetochore-organizing unit (Chapter III, Section C,4) and that the kinetochore itself is probably a temporary structure, disassembled at the end of one division and reassembled at the beginning of the next (see Chapter III, Section C,6,*a*). This mechanism might also find an analogy in the organization of meiotic chromosomes for synapsis, where apparently the face of the chromatid presented to its homolog across the synaptonemal complex is 90° from the face involved in pairing of sister chromatids (see Nebel and Coulon, 1962; Roth, 1966; Moens, 1968; see Moses, 1968, for review).

b. Rearrangement of Chromatin in Kinetochore Region

Alternatively, the same portions of the kinetochore region may be involved in both the first and second meiotic divisions, the change in orientation being brought about by means of a rearrangement or re-coiling of the underlying chromatin fibers. This possibility was considered by Östergren (1951) in his observation that in many organisms the chromosomes of the first meiotic division are more compact, have major coils of relatively large diameter, and show no constriction of the kinetochore region. He suggested that because of this coiling sister kinetochores in the first meiotic division are restricted to one side of the chromosome. Chromosomes of mitotic and second meiotic divisions, however, have a smaller diameter, a primary constriction, and lack major coils; therefore, the sister kinetochores are free to take up positions on opposite sides of the chromosome.

It is doubtful, however, that the presence or absence of primary constrictions and major coils is of prime importance in the orientation of sister kinetochores, because meiotic chromosomes with conspicuous primary constrictions and without major coils may show normal orientation at metaphase I (e.g., Dowrick, 1957; John and Lewis, 1965, Fig. 14). Furthermore, the observations of Schrader (1936) and Luykx (1965b) indicate that the reorientation of sister kinetochores between the first and second meiotic divisions takes place without the appearance of an obvious constriction in the kinetochore region.

The basic idea that sister kinetochore reorientation is the result of a rearrangement of the underlying chromatin is not disproved by these observations. A few workers have noted incidentally that meiotic cells treated with uncoiling agents such as metabolic inhibitors (Kemp, 1964) or high temperatures (Jain, 1957; Henderson, 1962) may sometimes show abnormal chromosome orientation. The differential coiling ("heteropycnosis") of the unpaired X chromosome in grasshopper spermatocytes is probably responsible, in part, for its differential kinetic behavior in the first meiotic division. Its unusual behavior cannot be ascribed solely to its unpaired condition, for, as Henderson (1962) noted, in cells in which unpaired autosomes are oriented at the equator of the spindle, with sister kinetochores facing opposite poles, the heteropycnotic X chromosome is still characteristically oriented toward one pole only. These observations do suggest a possible relation between the kind of coiling or folding of chromatin fibers in the kinetochore region and the disposition of sister kinetochores on the chromosome surface. The question deserves more careful and systematic study than it has yet received.

c. Traction Forces Acting from Opposite Poles

Traction forces acting on the sister kinetochores from opposite directions might also be an important factor in the change in their position from adjacent to opposite sides of the chromosome during meiosis. To be sure, a redisposition of sister kinetochores on the chromosome surface is not brought about by their transitory connection to opposite poles during prometaphase I (Luykx, 1965b); and certainly the slight separation that occurs during anaphase I (see Chapter III, Section D,1) is not caused this way. Such observations suggest that some internal change in the kinetochore region must take place before sister kinetochores can move to opposite sides of the chromosome. It is conceivable, however, that after the first meiotic division is over it is the chance connection of sister kinetochores to opposite poles in prometaphase II that is instrumental in bringing them to lie on opposite sides of the chromosome.

Some support for this idea can be found in the behavior of unpaired chromosomes in the first meiotic division.

d. Behavior of Univalents

Pairs of homologous (or partly homologous) chromosomes that are linked together (usually by one or more chiasmata) are known as *bivalents*; chromosomes that enter the first meiotic division in an unpaired condition are known as *univalents*. To distinguish them from univalents, the chromosomes that have disjoined from bivalents (called "tetrads" in some of the older literature) at anaphase I are generally referred to as *dyads*. The phrase *sister kinetochores* designates the two kinetochores of a univalent, of a dyad, or of one member of a bivalent.

Univalents are found in meiotic cells in which there are chromosomes without homologous partners, where normal pairing does not take place (asynapsis), or where homologous chromosomes do pair but fail to form chiasmata (desynapsis). Univalents occur in meiosis in haploid, triploid, and aneuploid organisms and in species hybrids. They can be induced in diploids by treating cells in early meiotic prophase with metabolic inhibitors or by subjecting them to abnormally high temperatures. Univalents also occur naturally in some organisms as unpaired sex chromosomes or as heterochromatic supernumerary chromosomes. For studies on chromosome orientation, the most instructive cases are those in which the behavior of univalents and bivalents can be observed in the same cell.

In general, where the degree of asynapsis or desynapsis is very high, the univalents are usually seen in two, often unequal, groups at the poles of the spindle; where few univalents are present, they often appear to be almost randomly distributed in the spindle—at the poles, at the equator, or somewhere between a pole and the equator (White, 1946; Henderson, 1962). On the basis of the recurrent pole-to-pole movements of unpaired sex chromosomes seen during prometaphase and metaphase in living cells (Dietz, 1956, 1969; Bauer *et al.*, 1961; Nicklas, 1961), it is likely that univalents are almost never passively distributed in the spindle. Except for the time they spend at or near the poles, they are probably almost always moving toward one of the poles or toward the equator. Why the average distribution of univalents in the spindle appears different in cells with many univalents, as opposed to cells with few univalents, is not clear. It may be related to variations in the relative amounts of time spent at a pole versus that spent traveling between poles; or to differences in behavior during early and late metaphase, with variation in the duration of metaphase.

Observations on living cells are required to provide more information on this question.

While there is considerable variation in the course of meiosis in cells with unpaired chromosomes (see Darlington, 1965, Ch. X), all univalents migrate between poles during prometaphase and metaphase or remain at the poles for various lengths of time. In general, they exhibit one of two types of behavior during the first meiotic division. In some cases, as for example in asynaptic and haploid *Triticum durum* (e.g., Martini and Bozzini, 1966; Lacadena and Ramos, 1968), in haploid *Pelargonium* (Daker, 1967), and in male grasshoppers (with an unpaired X chromosome), the univalent is distributed at random to one pole or the other at anaphase. In other cases the univalent takes up a position at the equator of the spindle, with its long axis at right angles to the spindle axis, and then divides at late anaphase or early telophase, with the daughter chromatids moving to opposite poles behind the dyads.

Where univalents are to become oriented at the equator during the first meiotic division, they usually do so during late prometaphase and metaphase, generally some time after congression of bivalents to the equator. Such behavior has been described in detail for the sex chromosomes in living spermatocytes of tipulid flies (Dietz, 1956, 1969; Bauer *et al.*, 1961). From fixed material, Matuszewski (1962) also has described, in the oocytes of a gall midge, a late congression of univalent germ line chromosomes to the equator. (These two cases are unusual in that the equatorial univalents remain undivided until the second meiotic division.) A gradual accumulation of univalents at the equator during metaphase, when bivalents have already oriented at the equator, has also been documented in the careful studies of Wagenaar (1961a,b) on pollen mother cells of *Triticum* hybrids.

Observations on fixed materials show that the two kinds of univalent behavior—poleward movement without division at metaphase or anaphase, and movement to the equator followed by division at late anaphase—may even occur in the same cell (e.g., Wagenaar, 1961b, Figs. 4C, 4F; Henderson, 1962, Figs. 26, 27; Newman, 1967, Figs. 13, 14, 19, 20; Lacadena and Ramos, 1968, Fig. 9). Apparently, most univalents not oriented at the equator at the beginning of anaphase move undivided to one pole or the other; univalents oriented at the equator usually remain there until late anaphase and then divide. Reorientation of univalents *during anaphase*, whether from polar to equatorial position or vice versa, may possibly occur occasionally

(Wagenaar, 1961b, Fig. 4C; Bauer *et al.*, 1961; John, 1957b) but is generally quite uncommon.

The difference in the anaphase behavior of dividing univalents on the one hand, and dyads segregating from bivalents on the other, might arise from an initial difference in the disposition of their sister kinetochores. The pairing process itself might affect the structure of the kinetochore region so that the unpaired chromosome first comes onto the spindle with sister kinetochores already partially or completely separated toward opposite sides of the chromosome. There are a number of observations that argue against this idea, however. These observations indicate that synapsis per se does not affect the kind of connections that sister kinetochores form with the poles and imply that when the spindle is formed and active chromosome movements begin the structure of the kinetochore region in univalents and in bivalents is the same.

(1) Chromosomes whose kinetochore regions have not undergone synapsis may nevertheless behave like members of bivalents if they are physically associated with another chromosome in some other way.

This appears to be the case in spermatocytes of *D. melanogaster*, in which the X and Y chromosomes are associated at metaphase I not by chiasmata but by the "stickiness" of special regions that Cooper (1964) has termed "collochores." The X and Y move to the metaphase plate along with the autosomal bivalents and usually segregate to opposite poles at anaphase I along with the autosomal dyads (see figures in Peacock and Erickson, 1965). In pollen mother cells of rye, the occasional nonspecific stickiness of two asynaptic univalents orients the univalents toward opposite poles (Östergren and Vigfusson, 1953). The X and Y chromosomes in the spermatocytes of the beetle *Adalia bipunctata* appear to be associated by the formation of a common nucleolus; in congression and segregation their behavior is usually the same as that of the autosomal bivalents (John and Lewis, 1960).

Particularly instructive in this regard is the behavior of a Y-autosomal translocation in spermatocytes of the crane fly *Tipula oleracea* (Bauer *et al.*, 1961). The normal Y chromosome always enters the first meiotic division as a univalent; it usually migrates from one pole to the other during prometaphase and early metaphase, reorients in late metaphase by developing spindle fiber connections to opposite poles, and moves to the equator, where it remains until the autosomal dyads have reached the poles (see Dietz, 1969). The translocation chromosome, consisting of the proximal portion of the Y (with its kinetochore) and the end of one of the arms of an autosome, behaves in the same way when it occurs as a univalent. When this same chromosome is part of a trivalent, however—its distal segment having formed a chiasma with the homologous autosome—its behavior is indistinguishable from the autosomal members of the trivalent; it is usually oriented toward only one pole or the other at the end of metaphase and segregates with the autosomal dyads at anaphase. The prophase history of the kinetochore region of the translocation chromosome is the same in all cases; there is therefore no basis for attributing the difference in its behavior in these different circumstances to an initial difference in its structure. Its behavior during metaphase and anaphase clearly depends only on whether or not it is physically associated with another chromosome.

(2) Chromosomes that are members of bivalents or trivalents, having undergone synapsis and chiasma formation, may nevertheless behave like univalents under certain circumstances.

They may change their orientation from one pole to the other, in a manner analogous to the pole-to-pole movements of univalents. This reversal of orientation has been observed in autosomal bivalents during prometaphase of normal division (Östergren, 1951; Bauer *et al.*, 1961; Dietz, 1969), in bivalents on a tripolar spindle (Dietz, 1959), in bivalents one of whose kinetochore regions has been irradiated with UV microbeam (Izutsu, 1961), in sex trivalents (Hughes-Schrader, 1943; Inamdar, 1949; Dietz, 1958), and in trivalents formed in individuals carrying a reciprocal translocation between an autosome and a sex chromosome (Bauer *et al.*, 1961).

Members of bivalents may also lag at the equator, with sister kinetochores apparently facing *opposite* poles at the start of anaphase. The two chromatids may separate completely during anaphase, with the sister kinetochores leading the way to *opposite* poles (Richardson, 1936; Klingstedt, 1939; White, 1961). This kind of behavior is very rare, of course, but it seems to be favored by configurations in which homologous kinetochore regions are widely separated in the bivalent, i.e., when they are joined only by a chiasma at the end of long chromosome arms. Although in these instances the homologous chromosomes have paired and formed chiasmata, the kinetochore regions are in effect more or less free to behave independently.

These examples show that the difference in behavior between univalents that lag at the equator and divide and dyads that are distributed undivided to the poles can be attributed entirely to whether or not the chromosome is physically associated with another chromosome in the early stages of movement. There is no evidence that the process of synapsis itself has any direct effect on the subsequent behavior of the chromosome. This implies that the structure of the kinetochore region is not affected by synapsis, and that at the beginning of active chromosome movement the structure of the kinetochore region in univalents and bivalents is the same. Yet by late anaphase a remarkable difference in behavior has appeared; dyads segregating from bivalents are oriented toward one pole or the other, with sister chromatids remaining associated in the kinetochore region, while univalents are dividing at the equator.

The equatorial position of some univalents in the first meiotic division, their orientation perpendicular to the spindle axis, the connection of sister kinetochores to opposite poles (see the observations of Forer, 1966; Dietz, 1969), the complete separation of daughter chromatids and their movement to opposite poles are all entirely normal features of chromosome behavior, yet in dyads segregating from bivalents at anaphase I, these events normally do not occur until the *second* meiotic division. The conclusion to be drawn is that they have been induced prematurely in univalents by means of traction fibers acting on sister kinetochores from opposite poles. It can be imagined that during prometaphase sister kinetochores may be connected to spindle poles in any one of a number of ways, depending on the disposition of the kinetochores and the chance orientation of the chromosome with respect to the poles (see Fig. 16). Some of these configurations are undoubtedly less stable than others, and as division progresses the chromosomes tend to be found with increasing frequency in the most stable orientation consistent with a particular disposition of sister kinetochores on the chromosome surface. Chance increases in the strength of the pull from both poles simultaneously during prometaphase, or the general increase in pulling force at the beginning of anaphase (see Chapter IV, Section B,4), may act on maloriented or reorienting univalents to bring sister kinetochores prematurely to face in opposite directions and thus lead to equatorial positioning and premature division.

That premature chromatid separation can be brought about by the action of the spindle on misaligned chromosomes is clearly demonstrated by the behavior of the sex chromosomes in the spermatocytes of *Mantis*

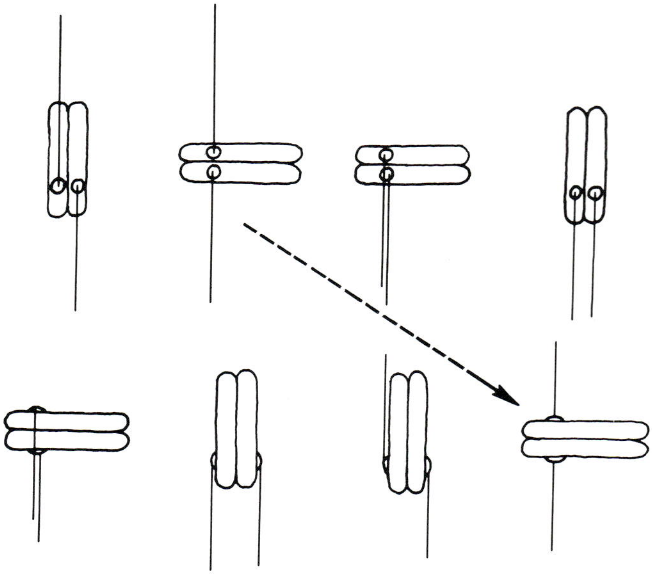

FIG. 16. Diagram showing various possible connections of sister kinetochores to opposite poles, with the sister kinetochores in the "adjacent" position in the first meiotic division (univalents, top row), in the "opposite" position in the second meiotic division (dyads, bottom row). The configuration on the right end of each row is the most stable for the two kinds of sister kinetochore positions. The configuration in the upper row, right end, is the most frequent orientation seen in univalents in prometaphase I and almost the only arrangement ever seen in dyads segregating from bivalents during anaphase I. The dashed arrow represents the transition presumed to occur in some univalents in the first meiotic division, leading to their division at the equator during anaphase or telophase I. The configuration in which sister kinetochores are located on opposite sides of the chromosome and are connected by spindle fibers to opposite poles (bottom row, right end) is the most stable of all; reorientation from this configuration is almost never observed (see Bauer *et al.*, 1961).

religiosa described by Callan and Jacobs (1957). In this material, X and Y chromosomes often remained connected by a relatively rigid chromatin bridge at the end of the first meiotic division. Because the spindles of sister cells in the second meiotic division are oriented at right angles to each other (and also perpendicular to the spindle axis of the preceding division), sister kinetochores at one end of the bridge oriented readily in the new spindle axis, while sister kinetochores at the other end of the bridge were subject to a torque that consistently caused them to separate prematurely.

The variable behavior expected from this model is consistent with what is actually observed; even within a single cell some univalents may be distributed undivided to the poles, while others divide at the equator. Even within these categories the variation is considerable; nondividing univalents may move to the poles during prometaphase, metaphase, or anaphase; the univalents dividing at the equator may do so at early anaphase or as late as telophase.

The behavior of univalents dividing in the first meiotic division strongly suggests that their sister kinetochores have come to lie on opposite sides of the chromosomes, as is the case with dyads in the second meiotic division. This is a prediction that in suitable material could easily be checked by means of electron microscopy. If true, it would mean that for chromosomes not stably oriented by metaphase I chance connections of sister kinetochores to opposite poles can prematurely bring sister kinetochores to face in opposite directions. This would therefore also be a plausible mechanism for the change in kinetochore disposition that normally occurs in dyads some time prior to metaphase II.

It is not known whether the final positioning of sister kinetochores on opposite sides of the dyad is in fact delayed until prometaphase II, as would be predicted if interaction with spindle fibers causes the change. Schrader was not able positively to identify kinetochores during interkinesis in *Amphiuma* spermatocytes, in spite of careful study (Schrader, 1936, 1939). Electron microscopy of serial sections through cells in which the critical stages can be unequivocally identified might give an answer. The question could also be approached experimentally by determining whether or not prevention of formation of the spindle for the second meiotic division (e.g., with colchicine or with low temperature) also prevents the change in the position of sister kinetochores from occurring.

The dyad segregating from a bivalent may be unable to reorient during anaphase I for a number of reasons. Its poleward orientation may be very stable, being maintained by its poleward movement. This possibility is supported by the observation that even in the prometaphase pole-to-pole movements of univalents most reorientations occur when the chromosome is *in the vicinity of the pole,* not when it is *moving between the poles* (Dietz, 1956; Nicklas, 1961; Bauer *et al.*, 1961). Second, the connections between kinetochores and spindle fibers may be firmer in anaphase than during earlier stages. This is suggested by the observation that reorientations of univalents *during anaphase* occur very rarely, if at all. Direct evidence for a firmer attachment of chromosomes to spindle fibers in anaphase is afforded by the micromanipulation studies of Nicklas and Staehly (1967). A third factor in maintaining the orientation of the dyad toward one pole during anaphse I might be the relative positions of the sister kinetochores themselves. Since the average bivalent achieves a stable orientation relatively early in division, sister kinetochores in the bivalent would be subjected to fewer reorientations and possible bipolar connections than would sister kinetochores in the univalent. This may allow sister kinetochores in dyads segregating from bivalents to maintain their adjacent positions throughout anaphase, so that the connection of sister kinetochores to *opposite* poles would be very unlikely, even if the anaphase movement and the spindle fibers were temporarily interrupted. The relative important of these various factors is a matter for experiment.

E. SUMMARY AND CONCLUSIONS

The kinetochore region of the chromosome in most eukaryotic cells is responsible for normal chromosome orientation and segregation during active stages of division and may contribute to the formation of spindle fibers, but it probably plays no essential role in other aspects of chromosome behavior at other stages in the cell's life cycle.

The kinetochore region consists of (1) the kinetochore organizer, a DNA-containing stretch of fibers probably containing many identical copies of the genes responsible for kinetochore function (Darlington's "centrogenes"); and (2) the kinetochore itself, a product of the activity of the genes in the kinetochore organizer. There is no concrete evidence that the kinetochore regions of the different chromosomes in a set are functionally different.

Sister kinetochores in chromosomes of the first meiotic division are situated adjacent to one another on the same side of the body of the chromosome. This arrangement is correlated with, and is largely responsible for, the reductional segregation of kinetochores that is characteristic of the first meiotic division. By the beginning of the second meiotic division, sister kinetochores have been relocated to opposite sides of the chromosome body in accordance with the separation of chromatids to opposite poles in that division. The mechanism by which this change in disposition of sister kinetochores between the first and second meiotic divisions is brought about is not known. Observations on kinetochore structure and on the behavior of unpaired chromosomes in meiosis suggest that internal changes in the kinetochore region and traction forces from opposite poles may both be involved, but further studies on the kinetochores of meiotic chromosomes are needed to answer this question.

CHAPTER IV

MECHANISMS AND MODELS OF CHROMOSOME BEHAVIOR

A. SPINDLE MICROTUBULES

Since microtubules are present in every dividing eukaryotic cell, from protist to man and angiosperm, that has been examined by adequate techniques, and since agents that disrupt spindle microtubules—e.g., low temperature, colchicine—also prevent chromosome movement or stop movement in progress, there is little doubt that microtubules are of major importance in normal chromosome movement.

1. Structure

The microtubules of the mitotic apparatus are recognizable in electron micrographs of sectioned material as long, relatively straight cylindrical structures with a diameter of about 250 A in cells fixed with glutaraldehyde, although they may appear as narrow as about 150 A in cells fixed with osmium tetroxide alone (see tabulation in Moor, 1967). In negatively stained preparations the lumen of the microtubule may be penetrated by the stain, indicating that the microtubule is indeed in some sense a hollow cylinder rather than a solid fiber. This is also shown in preparations from cells or isolated mitotic apparatuses spread on the surface of hypotonic media, where the microtubules are often split open lengthwise to form flat ribbons (Kiefer *et al.*, 1966; Barnicot, 1966). Such preparations reveal that the wall of the spindle microtubule consists of 10 to 13 longitudinal filaments, each composed of a string of globular particles 35–40 A in diameter. The longitudinal rows of particles are laterally associated in such a way as to give cross-striations with a pitch of 10–20°. These observations are in agreement with images of cross-sectioned spindle micro-

TABLE III

CORRELATION OF BIREFRINGENCE AND MICROTUBULES IN CELL DIVISION[a]

Material	Polarizing microscopy[b]	Electron microscopy[c]	References[d]
Haemanthus endosperm cells, prophase-prometaphase	Clear zone uniformly B, nucleus non-B; at time of nuclear envelope breakdown, B extends into nuclear area and becomes organized into streaks	MT's found only in clear zone during prophase; somewhat irregularly arranged; at time of nuclear envelope breakdown, MT's extend from clear zone into nuclear area in bundles	Inoué and Bajer, 1961 (PM); Bajer and Molè-Bajer, 1969 (EM)
Haemanthus endosperm cells, metaphase-anaphase	B visible as streaks extending from kinetochores toward polar regions	Bundles of MT's attached to kinetochores	Inoué and Bajer, 1961 (PM); Harris and Bajer, 1965 (PM, EM)
Haemanthus endosperm cells, late anaphase-telophase	B weak in interzonal region; increases as cell plate is formed; strongest at plate, weaker near daughter nuclei	MT's in interzonal region frequently in "crossed" arrangements; not parallel; as cell plate forms, MT's aggregate in parallel bundles; numbers of MT's increase during cell plate formation; few MT's near daughter nuclei	Inoué znd Bajer, 1961 (PM); Bajer, 1968c (EM); Hepler and Jackson, 1968 (EM)
Urechis egg, first cleavage division	Strong B in normal spindle, weak B in spindle area of artificially activated eggs	In normal spindle, MT's oriented approximately parallel; in activated eggs, MT's oriented in random directions	Luykx, unpublished

Sea urchin and clam eggs, isolated mitotic apparatus	B strong in asters and spindle, weak or absent in centrospheres	MT's oriented radially or approximately parallel in asters and spindle; oriented at random in centrosphere	Rebhun and Sander, 1967 (PM, EM); Harris, 1965 (EM)
Sea urchin eggs, isolated mitotic apparatus	Strong B in freshly isolated MA, gradual loss of B with storage	MT's abundant in freshly isolated MA; absent in stored MA	Goldman and Rebhun, 1969 (PM, EM); Kane and Forer, 1965 (PM, EM)
Pectinaria (annelid) oocyte	B increases on immersion in D_2O; B lost on treatment with colcemid	Increase in total number of MT's after D_2O treatment; loss of MT's after colcemid treatment	Inoué and Sato, 1967 (PM, EM)
Pectinaria oocyte	Gradual loss of B on treatment with vinblastine, griseofulvin; recovery on return to sea water	Gradual loss of MT's during treatment, reappearance of MT's after return to normal sea water	Malawista *et al.*, 1968 (PM, EM)
Chaetopterus (annelid) oocytes; lily pollen mother cells	Gradual loss of B at low temperatures, recovery on rewarming	—	Inoué, 1964 (PM)
Giant ameba	—	Greatly reduced number of MT's in cold cells, reappearance of MT's after rewarming	Roth, 1967 (EM)

[a] Observations with the polarizing microscope were made on living cells, except in cases in which the mitotic apparatus was isolated.

[b] B, birefringence or birefringent; MA, mitotic apparatus.

[c] MT, microtubule; MA, mitotic apparatus.

[d] PM, polarizing microscopy; EM, electron microscopy.

tubules in fixed material, where the wall of the microtubules is about 40–60 A thick and appears to be composed of 10 to 13 subunits (Behnke and Forer, 1966; Tucker, 1967). They agree also with Moor's observations on spindle microtubules in freeze-etched yeast cells, where similar subunits and a cross-striation with a pitch of 10–15° can be clearly seen (Moor, 1967). In their structure the spindle microtubules are thus very similar to the microtubules seen in other cellular locations (see Porter, 1966, for review) and undoubtedly share with them a number of properties, including the helical arrangement of subunits (Grimstone, 1967), lability (see Behnke and Forer, 1967; Gibbins *et al.*, 1969b), and rigidity (Pease, 1963; Wolfe, 1965). As would be expected for relatively rigid structures, the microtubules of the spindle follow straight (or sometimes gently curving) paths for distances of several microns. The waviness of spindle microtubules often seen in sections may be a shrinkage artifact occurring during dehydration of the specimen (Jensen and Bajer, 1969).

The 35- to 40-A globular subunit of spindle microtubules may correspond to a protein that has been extracted from the mitotic apparatus of the sea urchin egg, a protein with a sedimentation coefficient of about 3.5 S and a molecular weight of about 68,000, possibly composed of two monomers linked by a disulfide bond. It contains bound nucleotide and is insoluble in neutral solutions containing calcium ion (Sakai, 1966; Mazia, 1967; Bibring and Baxandall, 1968). A microtubular protein with similar characteristics has been extracted from cilia and flagella (Stephens, 1968; Renaud *et al.*, 1968); this protein is similar in a number of its properties, including amino acid composition, to the muscle protein actin (Stephens, 1968; Mohri, 1968).

2. Distribution and Classification

The distribution and numbers of oriented microtubules seen in electron micrographs generally correspond well with the distribution and intensity of birefringence seen in living cells by means of polarizing microscopy. Table III documents this assertion in a variety of cell types, showing that changes and differences in birefringence in normal cells and in cells treated in various ways are very closely correlated with changes in the numbers and orientation of microtubules. Similar correlations between birefringence and the presence of microtubules have been observed in heliozoan axopods (Tilney and Porter, 1967; Tilney, 1968). While the spindle may contain other components that give rise to birefringence or that modify its intensity (see Rebhun and Sander, 1967; Goldman and Rebhun, 1969), these observations support the idea that the degree of birefringence observed in living, dividing cells is primarily a function of the number and disposition of microtubules. On this basis, the rapid and reversible changes in spindle birefringence seen during brief drug treatments, chilling, and in normal cells especially in prometaphase, can justifiably be interpreted to mean that the spindle microtubules are in a dynamic state, continuously being broken down and formed anew throughout most of division (Inoué and Sato, 1967).

Polarizing microscopy of living cells and electron microscopy indicate

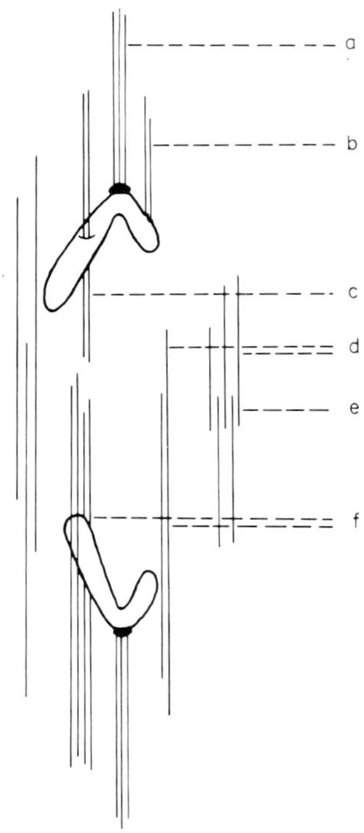

FIG. 17. Classification of spindle fibers. (a) Kinetochore fiber; (b) neocentric fiber; (c) penetrating fiber; (d) interzonal fibers; (e) fibers of midbody or phragmoplast; (f) sheath fibers; (a, b, and c) are chromosomal fibers; (c, d, e, and f) are continuous fibers.

that most of the microtubules are not uniformly distributed throughout the spindle but are loosely associated in sheets and bundles—the spindle "fibers" of light microscopy. A classification of spindle fibers, based on their locations in the spindle, is given in Fig. 17. The terminology is primarily that used by Bajer (1968b) and Inoué and Sato (1967) and is based on common usage. The following additional points should be noted.

The term "kinetochore fiber" is synonymous with the older term "chromosomal fiber." The presence of other spindle fibers closely associated with the chromosome—neocentric fibers (Fig. 7) and penetrating fibers (Fig. 13)—make the distinction necessary. With a few exceptions (see below), neocentric and penetrating fibers occur only sporadically. Penetrating fibers could be called either continuous fibers or chromosomal fibers. On the one hand, they pass through the chromosome without interruption and presumably make up part of the population of interzonal microtubules at anaphase (see Robbins and Gonatas, 1964, Fig. 16; Behnke and Forer, 1966; Buck, 1967; Jokelainen, 1967, Fig. 21). On the other hand, they are indeed closely associated with the chromosomes and may (as Behnke and Forer have suggested for crane fly spermatocytes) play an essential role in chromosome movement in some cell types. Sheath fibers, distinct bundles or sheets of continuous microtubules that surround the individual chromosomes, have been clearly seen in *Haemanthus* endosperm cells by means of electron microscopy (Bajer, 1968b); observations of a similar kind have been made on crane fly spermatocytes and lily pollen mother cells by means of polarizing microscopy of the living cells (Dietz, 1963; Inoué and Sato, 1967). Since the microtubules of these fibers often run parallel and very close to the kinetochore microtubules (e.g., P. Harris, 1962b, Fig. 4; Luykx, 1965b, Fig. 2; Bajer, 1968a, Plates 7 and 13), they may also play an important role in chromosome movement (see Chapter IV, Section C,5).

With respect to electron density and diameter, seldom are there obvious morphological differences among the microtubules located in different regions of the spindle or between spindle microtubules and those found in the centrosphere and aster. Likewise, no differences, other than changes in length, have been seen in microtubules at different stages of division. On the basis of changes in the distribution of spindle birefringence during the normal course of division, Inoué suggests that the components of spindle fibers in one location may be disassembled and then reassembled to form spindle fibers at another location, the changes being determined by the successive activation and deactivation of orienting centers (Inoué, 1964; Inoué and Sato, 1967). Electron microscope observations on changes in the numbers and distribution of microtubules have been given similar interpretations (e.g., Bajer, 1968b; Hepler and Jackson, 1968).

The possibility that there are structural differences in microtubules at different locations in the spindle or structural changes in spindle microtubules during the course of division can by no means be discounted, however. First, the apparent diameter of microtubules in different materials under different conditions of preparation is notoriously variable (see Moor, 1967); the electron density of microtubules might also be expected to vary with section thickness and with fixation and staining procedures. Such variability could easily obscure any small differences that might exist. Second, few systematic measurements on spindle microtubules in different locations or at different stages have been attempted. Allen and Bowen (1966) did note the presence of microtubules of two size classes—180 and 280 A in diameter—in sporogenous cells of *Psilotum nudum* in prophase; in later division stages only the larger microtubules were visible. And in dividing yeast cells Moor (1967) found that microtubules extending outward from the pole into the cytoplasm averaged about 250 A in diameter, while those extending from the pole into the nucleus had an average diameter of about 224 A; there was no noticeable change during spindle elongation. Third, several observations suggest that kinetochore fibers are less sensitive to chemical and physical treatments than are other fibers of the mitotic apparatus. Brief exposure of dividing sea urchin eggs to colcemid disorganizes the asters, but kinetochore fibers remain (Sauaia and Mazia, 1961). Similarly, in Chinese hamster cells exposed to a low concentration of colcemid continuous spindle microtubules were absent and centriole separation was prevented, while some kinetochore microtubules remained (Brinkley *et al.*, 1967). In a preliminary observation Behnke and Forer (1967) treated crane fly spermatocytes with colchicine and found that the chromosomal microtubules seemed to persist longer than other spindle microtubules. Exposure of *Haemanthus* endosperm cells to chloral hydrate seems to affect primarily the development of continuous fibers; in spite of the overall disorganization of the spindle, bundles of microtubules attached to the kinetochores can still be found (Molè-Bajer, 1969). Inoué (1964) observed that after chilling lily pollen mother cells in anaphase the kinetochore fibers were the first to recover their birefringence. Specht (1961) has reviewed a number of observations from older literature that also suggest differences in sensitivity to inhibitors among spindle fibers in different locations.

In short, in the size differences that have been seen in yeast and in *Psilotum*, and in the differences in response to spindle inhibitors, there is some basis for thinking that not all spindle microtubules are functionally alike. In particular, chromosomal microtubules are less sensitive to the effects of colchicine, chloral hydrate, and cold than are continuous microtubules. The observation of Ris (1949) that chloral hydrate prevents spindle elongation but does not inhibit movement of the chromosomes to the poles is consistent with these observations. It also suggests that continuous fibers are not essential for movement of the chromosomes to the poles but do play a direct role in spindle elongation. Elongation or shortening of spindle fibers during division does not appear to be accompanied by any other structural changes in the microtubules, however.

Since acentric fragments do not behave normally in division, attention is naturally drawn to the kinetochore fibers for an explanation of chromosome movements. Carlson (1938) observed, however, that even the two daughter chromatids of acentric fragments, induced by X-irradiation in grasshopper neuroblasts, moved toward opposite poles in late anaphase more often than they moved to the same pole. This observation suggests that spindle microtubules other than those attached to kinetochores may also play important roles in chromosome movement. In some cells chromosome separation is accomplished by spindle elongation alone, with little or no shortening of kinetochore fibers (see Mazia, 1961, pp. 272–279). As mentioned earlier (Chapter III, Section C, 6,*d*), some recent electron microscope studies have failed to reveal definite kinetochore fibers in some cell types; such findings place the burden of chromosome movement in these cells on spindle fibers in other locations.

In cells in which the spindle is not too long to allow a sufficiently high proportion of favorably oriented sections, single microtubules can occasionally be seen to traverse without interruption the whole distance from pole to metaphase chromosome. This appears to be true for vertebrate cells as well as for microorganisms, and holds for cells in which the chromosomes approach the poles at anaphase as well as for cells

TABLE IV

OBSERVATIONS OF UNINTERRUPTED SINGLE MICROTUBULES BETWEEN CHROMOSOME AND POLE

Material	Chromosome-to-pole distance (μ)		Spindle length (μ)		Reference
	Metaphase	Late anaphase	Metaphase	Late anaphase	
Plasmodium (malarial parasite)	0.5	?	1.2	?	Aikawa and Beaudoin, 1968, Figs. 3, 4.
Nassula (ciliate) micronucleus	2–3	<2	6	100	Tucker, 1967, Fig. 4.
Physarum (slime mold) plasmodium and myxameba	2–4	0.5–1.0	6–7	10	Aldrich, 1969, Figs. 3, 9.
Coprinus (basidiomycete) meiosis	1.5	0.1	2–3	?	Lu, 1967, Figs. 22, 23.
Catenaria (water mold)	1.5	<0.1	3	4–5	Ichida and Fuller, 1968, Fig. 8
Rat embryo cells	2	—	4–6	—	Jokelainen, 1967, Figs. 7, 8, 15.
Chick mesenchymal cells	2.5–3.0	0.5	5–6	8–10	Allenspach and Roth, 1967, Fig. 7

in which anaphase separation is accomplished mainly by spindle elongation (Table IV). Thus, it would appear unnecessary to postulate a staggered arrangement of laterally associated, shorter microtubules spanning the distance between kinetochore and pole. It seems likely that the kinetochore fiber consists of microtubules many of which individually traverse the whole distance. This observation is incompatible with schemes in which kinetochore microtubules of constant length slide past one another as the kinetochore fiber shortens in anaphase. Such a mechanism would demand that the number of microtubules at a given point in a kinetochore fiber increases during poleward movement and that the anaphase chromosome does not approach the pole more closely than the length of individual microtubules attached to the kinetochore. Yet Bajer (1968b) has found in *Haemanthus* endosperm cells that the number of kinetochore microtubules may *decrease* during anaphase and, as the data in Table IV show, late anaphase chromosomes in many cells are much closer to the pole than the length of individual microtubules attached to the metaphase kinetochore would allow. For cells in which the chromosomes approach the poles during anaphase, then, the individual microtubules comprising the shortening kinetochore fiber must themselves shorten.

Some of the microtubules of the continuous fibers run the entire distance from pole to pole. In many cells microtubules can be seen passing through the equatorial region without interruption (e.g., Roth *et al.*, 1966, Fig. 2; Bajer, 1968b, Fig. 7), but how far such microtubules extend poleward in either direction usually cannot be determined. Individual microtubules connecting the two poles can be seen in two published micrographs: in the division figure of the malarial parasite *Plasmodium gallinaceum,* in which the pole-to-pole distance is only 1.2 μ (Aikawa and Beaudoin, 1968, Fig. 4), and in the short early meiotic prophase spindle in the diatom *Lithodesmium undulatum* (Manton *et al.*, 1969, Fig. 23). In the latter case, however, *most* microtubules extend from one pole only about three-quarters of the way to the opposite pole; this is apparent in longitudinal sections, and cross sections of the spindle show approximately twice as many cross-sectioned microtubules in the equatorial region as in the regions close to either pole. Manton and her co-workers have estimated that only about 10% of the continuous microtubules actually extend the entire distance from pole to pole. A similar arrangement can also be found in mammalian cells and in higher plant cells, at least in late anaphase and telophase. While a few individual microtubules in these late stages may pass without interruption through the equatorial region (Robbins and Gonatas, 1964, Fig. 24; Bajer, 1968c, Fig. 10), most of the interzonal microtubules terminate in the midbody or phragmoplast (Pawaletz, 1967; Byers and Abramson, 1968; Hepler and Jackson, 1968). Thus, in a real sense the spindle may consist of two half-spindles.

B. GENERAL FEATURES OF MITOSIS

Any model of chromosome movement and orientation must take account of the following generally valid conclusions.

1. Attachment of Chromosomes to Spindle

From the beginning of prometaphase to the end of anaphase, chromosomes are firmly attached to spindle fibers. This has been most convincingly demonstrated in the micromanipulation experiments of Nicklas and Staehly (1967). Meiotic chromosomes in grasshopper spermatocytes can be stretched with a microneedle during prometaphase, metaphase, and anaphase without causing much change in the distance between kinetochore and pole. Bivalents and their associated kinetochore fibers (observed by polarizing microscopy) can easily be moved as a unit, however, in a direction transverse to the spindle axis, from one side of the spindle to the other. These observations indicate that the chromosome is firmly attached by its kinetochores to the individual kinetochore fibers. The ease of lateral displacement of a bivalent with its kinetochore fibers also suggests that the kinetochore fibers themselves are *not* firmly anchored in the spindle. Curiously, the kinetochore of the anaphase chromosome resists being moved away from the pole toward which it is oriented but is easily displaced poleward and then back again to its original anaphase positions.

2. Independence of Chromosome Movements

From the beginning of prometaphase to the end of anaphase, chromosomes usually move as individuals. Therefore, in seeking explanations for chromosome movement it is necessary to think in terms of localized forces that act on individual spindle fibers or on individual chromosomes. Sometimes the direction or rate of movement of one chromosome can be influenced, or even completely determined, by the behavior of another chromosome (see Chapter IV, Sections E,2 and 3); such cases are by no means unimportant, but the effects are either slight or are clearly special modifications of the normal mitotic process.

The independence of the movements of individual chromosomes is clearly observed in living cells during the establishment of the metaphase plate (Dietz, 1956; Bajer and Molè-Bajer, 1956; Bajer, 1958c; Nicklas, 1965). Chromosomes may take very different paths to reach the equator and may arrive at different times. Once at the equator, they oscillate independently in the long axis of the spindle. As already noted (Chapter III, Section D,2,*d*), unpaired chromosomes in the first meiotic division may migrate from pole to pole during prometaphase and metaphase, may move to the poles in advance of the dyads at anaphase, or may lag at the equator as the anaphase dyads move poleward. Bajer (1968a) has observed that in highly flattened *Haemanthus* endosperm cells at early anaphase two chromatids from different chromosomes may pass each other moving in opposite directions on their way to the poles. Extensive micromanipulation of individual chromosomes has no observable effect on the behavior of other chromosomes in the same cell (Nicklas and Staehly, 1967; Nicklas, 1967).

3. Pulling (Traction) Forces

The force acting on the kinetochore region during all stages of active chromosome movement is best described as a *pulling* force, that is, the force is applied to the kinetochore region from the same direction in which the chromosome moves.

Occasionally in early prometaphase of mitosis, as a chromosome first begins to move rapidly, one of the two sister kinetochores is stretched poleward more than the other; chromosome movement is in the direction of the stretched kinetochore (Bajer, 1958b; Bajer and Molè-Bajer, 1963). Dietz (1956, 1969) has made similar observations in prometaphase of the first meiotic division in crane fly spermatocytes. Such observations are most simply explained as being caused by a pulling force acting on the stretched kinetochore.

Most of the structures implicated in the movement of chromosomes are found on the *poleward* side of the chromosome. The kinetochore itself, as seen in electron micrographs, is located on the poleward side of the chromosome body during metaphase and anaphase. Most of the birefringence and most of the microtubules of the spindle are found on the poleward side of the chromosomes during anaphase. UV microbeam irradiation of the region between the anaphase chromosome and the nearest pole is more effective in stopping chromosome movement than is irradiation of the interzonal region (Forer, 1966).

A convincing demonstration that the equatorial position at metaphase is maintained by antagonistic pulling forces acting on opposite kinetochores is found in the work of Izutsu (1961) on grasshopper spermatocytes. Izutsu observed that irradiation of one kinetochore pair of a bivalent or the adjacent kinetochore fiber caused the bivalent to shift in the direction of the pole toward which its *unirradiated* kinetochore was oriented. From the observation that anaphase chromosome movement is stopped or prevented by the irradiation, it is clear that the effect of the UV microbeam is to disrupt, not to activate, the mechanism responsible for chromosome movement. On this basis, the observed shift of the metaphase bivalent must mean that it is subject to *pulling* forces acting on the two kinetochore pairs. If the position of the bivalent at the equator were maintained by repulsion from opposite poles, the direction of the movement after irradiation would be opposite to that actually observed.

Finally, as Östergren (1950, 1951) has emphasized, the metaphase orientation of chromosomes in mitosis and meiosis, in such a way that sister kinetochores or homologous kinetochore pairs are lined up in the long axis of the spindle, is very difficult to understand on the basis of pushing forces acting away from the poles but is readily explained by assuming that kinetochores are subject to poleward pulling forces throughout prometaphase.

As noted in Chapter III, Section A and Chapter IV, Section B,1, kinetochores are attached to spindle fibers at all stages of active chromosome movement. This observation, along with those presented above, suggests that the mechanism responsible for prometaphase movement and establishment of the metaphase plate is fundamentally the same as that responsible for movement of the chromosomes to the pole at anaphase, and that it is based on a pulling force directed toward the poles.

In the divisions of primary spermatocytes in *Sciara* (Metz *et al.*, 1926) and in the meiotic divisions of the "harlequin" lobe of the testis in certain hemipteran insects (Schrader, 1946a,b), some chromosomes move away from the pole toward which their chromosomal fibers extend. In these exceptional cases other factors probably override the normal pulling force (Chapter IV, Section C,6).

4. Variations in Pulling Forces

The poleward-directed force acting on the kinetochore varies in magnitude during the course of division.

The average force acting on the individual kinetochore is large during prometaphase, decreases gradually as the cell progresses toward metaphase, and then increases abruptly at the onset of anaphase. This conclusion is based mainly on observations of the degree of stretching of chromosome segments lying between opposed kinetochores and on the position of multivalents in the spindle, but it is supported also by observations on kinetochore fiber birefringence and rates of chromosome movement.

Early prometaphase of the first meiotic division in oocytes and spermatocytes of animals from widely separated phyla (see tabulation in John, 1957a) is often characterized by a "premetaphase stretch." The attenuation of the chromosome in this process cannot be caused by an active elongation, because in the bivalent it is seen only in the chromosome segment that connects the two homologous kinetochore pairs. It is seen neither in regions of bivalents distal to the most proximal chiasmata nor in univalents. It cannot be the result of repulsion between kinetochore pairs, because it is seen neither in maloriented bivalents (Henderson, 1965, Fig. 11), nor in segments of multivalents that connect kinetochore pairs oriented toward the same pole (Staiger, 1955). It must therefore be the result of a pulling force acting from the poles on the opposed kinetochore pairs. Part of the stretching can be accounted for by the elongation of the spindle during prometaphase. However, the separation of kinetochores during stretch is sometimes greater than the elongation of the spindle. It may also occur at different times for the different bivalents in a cell, so that the stretching of some bivalents takes place after spindle elongation is completed (Hughes-Schrader, 1943). In some species it may take place without any change in spindle length (Dietz, 1958). These observations mean that during the prometaphase stretch, as in anaphase, the kinetochore fibers may shorten and the kinetochores move closer to the poles.

As prometaphase progresses, the degree of stretching decreases and homologous kinetochore pairs approach each other. Again, this shortening of bivalents is sometimes accompanied by a shortening of the spindle as a whole, but in other instances it occurs without appreciable change in spindle length (Dietz, 1958).

Such changes are not limited to the meiotic cells of animals. In the mitotic chromosomes of *Haemanthus* endosperm, stretching of kinetochore regions occurs in early prometaphase but decreases in extent and frequency as the cell progresses toward metaphase (Bajer and Molè-Bajer, 1963).

Such observations indicate that the magnitude of the poleward forces acting on kinetochores is decreasing as the cell progresses toward metaphase. By late metaphase the poleward forces are at a minimum, or in some cells may even be absent altogether (see below).

A major alternative to be considered is the possibility that the gradual shortening of the bivalent during prometaphase reflects a condensation or coiling of the chromosome or a change in its elastic properties, resulting in a contraction that gradually overcomes the opposed poleward pulling forces on the kinetochores. The shortening of stretched bivalents as prometaphase progresses has often been explained this way. This interpretation implies, however, that there is tension in the chromosome at metaphase sufficient to oppose the degree of stretching observed earlier, and is difficult to reconcile with the observation of Nicklas and Staehly (1967) that metaphase bivalents in grasshopper spermatocytes can be stretched with a microneedle to several times their original length without much change in the position of their kinetochores.

The supposition that kinetochores are subject to pulling forces of the same magnitude at metaphase as they were in earlier stages is incompatible with other observations also. In many cases chromosomes connected by spindle fibers to only one pole at the end of metaphase do not move immediately to that pole but wait for the onset of anaphase. This kind of behavior was first described by Hughes-Schrader (1948a) for the X chromosome in spermatocytes of the mantid *Humbertiella indica*. It has been observed in chromatids prematurely separated by micromanipulation (Nicklas, 1967); it is also seen during normal division in cells in which chromatid separation slightly precedes anaphase movement (Schrader, 1936; Molè-Bajer, 1958; Dietz, 1969).

The position of trivalents during the first meiotic division also indicates that the poleward forces acting on kinetochores decrease to a minimum at metaphase (see review and discussion by Dietz, 1958). During prometaphase the trivalent generally lies off the equator, closer to the pole toward which two of its three kinetochores are oriented (see, for example, the figures in Staiger, 1954, 1955). This observation is consistent with the notions that the equilibrium position of the chromosome reflects the initial difference in the forces pulling kinetochores toward opposite poles and that the forces acting on different kinetochores are additive. By metaphase, however, the trivalent usually lies close to the equator.

Another alternative that might be proposed to account for all of these characteristics of metaphase (a reduction in the attenuation of prometaphase bivalents, the equatorial position of trivalents, and the equatorial position of chromosomes with kinetochore fibers to only one pole) is that the pulling force on the kinetochores is opposed by a gelation spreading from the poles toward the equator of the spindle, so that even though the pulling force might remain large through metaphase its effects on stretching and chromosome position are overcome by an increasing rigidity of the region between the pole and the equator. This possibility can be excluded on the basis of observations on the movements of univalents, chromosome fragments, persisting nucleoli, small lipid droplets, and so on, in this region throughout prometaphase and metaphase (see Chapter IV, Section B,6).

The time that the univalent X chromosome of grasshopper spermatocytes spends in motion between the two poles is somewhat greater in early prometaphase than in late prometaphase (Nicklas, 1961). The kinetochore fibers of *Haemanthus* endosperm cells may be less birefringent at metaphase than during prometaphase (Bajer and Molè-Bajer, 1963). These two observations provide additional, although indirect, support for the conclusion that the poleward forces acting on kinetochores at metaphase are a minimum.

In some cell types the magnitude of the pulling forces seems to remain at a high level until the onset of anaphase. Izutsu's (1961) observation that the irradiation of one kinetochore pair results in a shift of the metaphase bivalent to the opposite pole indicates that in these cells (primary spermatocytes in several species of grasshopper) poleward forces were still acting on the kinetochores at the time of irradiation. Bivalents in spermatocytes of the salamander *Batrachoseps wrighti* exhibit stretching and oscillations up to the beginning of anaphase (Seto *et al.*, 1969). Unlike the bivalents of spermatocytes of other crane flies, those of *Tipula oleracea* show continued stretching until anaphase, and trivalents in this species remain off the equator, closer to the pole toward which two of the kinetochore pairs are oriented (Bauer *et al.*, 1961; Dietz, 1969). Dietz has interpreted this to mean that in these cells anaphase begins before full "metaphase" (as judged by decreased stretching, equatorial position of trivalents) is attained.

At the beginning of anaphase, the poleward forces acting on kinetochores increase suddenly. This is shown in the pronounced stretching of kinetochore regions that fail to separate or are delayed in their separation [in spermatogonia of grasshopper species hybrids (Klingstedt, 1938, 1939); in chromosomes whose kinetochore region has been irradiated with a UV microbeam (Bajer and Molè-Bajer, 1961); in the "early anaphase stretch" of *Chortophaga* spermatocytes (Nicklas, 1963).] Again, it is not likely that the poleward stretching of the kinetochores in these cases is due solely to a change in the extensibility of the chromosome. Independent evidence for an increase in the poleward pull at anaphase is also found in the following observations: (1) metaphase chromosomes with kinetochore fibers to only one pole that are maintaining their position some distance away from that pole suddenly move poleward (e.g., Hughes-Schrader, 1948a; Nicklas, 1967; see also Östergren and Bajer, 1961; Östergren, 1961); and (2) trivalents that are positioned at the equator at metaphase, and that fail to divide, move off the equator toward the pole faced by two of the three kinetochore pairs (Dietz, 1958). Correlated with this evidence for an increased pull on the kinetochores is the observation that the birefringence of kinetochore fibers becomes strongest just at the onset of anaphase (Inoué and Bajer, 1961; Inoué, 1964).

Superimposed on these changes are small, relatively rapid fluctuations in the strength of the pull on the kinetochores, as indicated by low-amplitude oscillations of the kinetochores in prometaphase and metaphase and slight irregularities in anaphase movement (e.g., Dietz, 1956; Bauer *et al.*, 1961; Nicklas, 1965; Bajer, 1968a; Seto *et al.*, 1969) and by fluctuations in the birefringence of kinetochore fibers (Inoué, 1964).

5. Two Kinds of Anaphase Movement

The separation of chromosomes at anaphase is accomplished by a shortening of kinetochore fibers, by an elongation of continuous fibers, or by both kinds of movement together (see discussion and references in Mazia, 1961).

6. Nonspecific Poleward Force

Throughout division there is a nonspecific force acting to move microscopically visible particles in a poleward direction. A wide variety of particles are moved poleward in a wide variety of cell types: chromosome fragments in grasshopper neuroblasts (Carlson, 1938); small unidentified particles and chromosome fragments in *Haemanthus* endosperm (Bajer and Molè-Bajer, 1956, 1963; Bajer, 1958a,b, 1967); heterochromatic chromatin granules in early cleavage stages of copepod embryos (Stich, 1954; Beermann, 1959); persistent nucleoli in meristematic cells of various plants (e.g., Frew and Bowen, 1929; Fabbri, 1960; Underbrink *et al.*, 1967); lipid droplets in newt heart cells in culture (Bloom *et al.*, 1955); fragments of the nuclear envelope in *Labyrinthula* (Perkins and Amon, 1969); and mitochondria, pigment granules, yolk, and other particles in the eggs of a wide variety of marine invertebrates (see reviews by Rebhun, 1959, 1960, 1963). In all of these cells the poleward movement depends on the presence of an organized spindle apparatus. As Rebhun (1964, 1967) has pointed out, the wide variety of different kinds of particles involved in such movements, among other observations, suggests that they are not self-propelled. The irregularity and independence of the movements of small particles, as well as the observation that the particles often move from all directions toward the pole (Östergren *et al.*, 1960; see Fig. 2), indicate that they are not carried along by a general poleward streaming of cytoplasm.

Detailed observations on the poleward movement of particles in living cells are of interest. The motion of small granules may be described as "saltatory;" periods of irregular, randomly directed movements alternate with sudden movements over a distance of several microns. Such movements also occur during interphase, but during spindle formation they occur more frequently, are more rapid, and tend to become oriented in the long axis of the spindle or radial to the centrosphere in the aster, with the frequency, rate, and distance of movement *toward* the pole tending to be greater than that *away* from the pole (see the observations of Bajer and Molè-Bajer, 1956; Bajer, 1958b, 1967, on particle movement during mitosis in endosperm; and those of Rebhun, 1967, in marine invertebrates and in newt cells).

The velocity of these movements can be very great, up to several microns per second, but their duration is rather short, being of the order of a few seconds. Averaged over longer periods of time, the net poleward movement would be considerably slower. The forces causing these movements must be extremely localized, for two particles less than 1 μ apart may separate as one of them undergoes saltation (Rebhun, 1967). These two considerations readily account for the observation that larger bodies such as acentric chromosome fragments move much more slowly and smoothly toward the poles, in fact, at a rate almost identical to that of chromosome movement in normal anaphase (Bajer and Molè-Bajer, 1963).

The relevance of these findings to the problem of chromosome movement is obvious. The force that moves various particles, nucleoli, and chromosome fragments poleward is probably identical to the force responsible for normal chromosome movements, as the following characteristics demonstrate:

(1) The force is generally directed toward the poles, as are most

chromosome movements; (2) it is active throughout the period of active chromosome movements; granules, chromosome fragments, and so on, are moved poleward during prometaphase and metaphase as well as anaphase; (3) it is applied locally throughout the spindle (as the independence of individual particle movements shows) and thus could account for the independence of the movements of individual chromosomes (Chapter IV, Section B,2); (4) on the average, over long periods of time, it moves particles poleward at approximately the same rates as that of normal chromosome movement; and (5) it may increase around the time of the transition from metaphase to anaphase; Bajer (1967) noted an increase in the rate of poleward movement of granules in *Haemanthus* endosperm cells at this stage. This observation is in accord with the evidence that the pulling force on kinetochores increases at the onset of anaphase (Chapter IV, Section B,4).

7. Noncontractility of Kinetochore Microtubules

Kinetochore fibers are probably not contractile, at least not along their whole length. During anaphase they appear to move poleward at the same rate as the chromosomes.

These conclusions are suggested by the observation that regions of reduced birefringence caused by UV microbeam irradiation of kinetochore fibers in anaphase often move poleward at approximately the same rate as the anaphase chromosomes (Forer, 1965, 1966). The length of the birefringent spindle fiber between the kinetochore and the region of reduced birefringence does not decrease, as might be expected if the whole fiber contracted during anaphase movement.

As Bajer (1966b, 1968a) noted, the same conclusion might be drawn from the behavior of certain chromosomes in *Haemanthus* endosperm cells that have been highly flattened or treated with chloral hydrate. Occasionally under these circumstances the daughter chromatids separate from two parallel metaphase chromosomes in such a way that two nonsister chromatids move poleward, one a short distance directly behind the other. In spite of the distance between them (2–3 μ), the two chromatids often appear to travel together as a unit, undergoing the same irregularities of motion, as if their kinetochore fibers were fused. For practically all of their poleward movement, however, the distance between the chromatids does not change, suggesting that the kinetochore fiber of the following chromatid, at least up to the level of the one ahead, does not contract during anaphase. Observations on birefringent kinetochore fibers in anaphase and on the movement of granules in the area of the kinetochore fiber also suggest that the kinetochore fiber is shifted poleward as a unit during anaphase (Bajer, 1961, 1967, 1968b).

These observations do not exclude the possibility that the kinetochore fiber is contractile only at its polar end. The presumption that during poleward movement at anaphase the microtubules of the kinetochore fiber do not contract at all is, however, consistent with the purely structural role that has been postulated for microtubules in other cellular locations (see Porter, 1966; Gibbins *et al.*, 1969a). Recent work indicates that the microtubules of cilia also do not contract during ciliary motion (Satir, 1967, 1968).

If kinetochore microtubules do not contract, then their shortening during anaphase must be accompanied by gradual disassembly. The simplest hypothesis consistent with the observations of Forer (1965, 1966) and Bajer (1966b, 1968a) is that the kinetochore microtubules are disassembled at their *polar* ends during poleward migration of the chromosomes at anaphase. Some support for this idea can be found in the progressively more irregular arrangement of microtubules in the polar region and the enlargement of the centrosphere during anaphase (Bajer, 1968a,b).

It might also be noted that models of anaphase movement involving a kinetochore fiber that is contractile require that the fiber be somehow anchored at the polar end. This assumption is not easily reconciled with the observation that anaphase movement can also occur without organized spindle poles (Chapter IV, Section B,11).

8. The Force-Generating Component of the Spindle

The force-generating elements of the mitotic apparatus represent a component distinct from the kinetochore microtubules.

Forer (1966) found that UV microbeam irradiation of kinetochore fibers at anaphase sometimes altered their birefringence without affecting poleward movement and at other times affected movement without altering birefringence. Furthermore, irradiations that did not change the birefringence of the kinetochore fiber were as effective in slowing or stopping chromosome movement as irradiations that did cause a change in birefringence (Table V). Forer also found that the effectiveness of irradiation in causing changes in *birefringence* appeared to vary with the wavelengths he used, while the effectiveness of irradiation in

TABLE V

ULTRAVIOLET IRRADIATION OF KINETOCHORE FIBERS AT ANAPHASE IN SPERMATOCYTES OF THE CRANE FLY *Nephrotoma suturalis*[a]

Anaphase movement	Number of irradiated kinetochore fibers with	
	Birefringence unchanged	Birefringence reduced
Slowed or stopped	87 (72%)	36 (71%)
Unchanged	34 (28%)	15 (29%)

[a] Data from Forer, 1966, Figs. 2, 3.

causing changes in *movement* did not vary with wavelength. The effects of UV irradiation on movement and on birefringence therefore seem to be entirely independent, suggesting that movement and birefringence are caused by two different spindle components.

Since birefringence is correlated with the presence of oriented microtubules (Table III, Chapter IV, Section A,2) and since it is known that UV irradiation affects the structure of microtubules (Bajer, 1969), it seems likely that the microtubules represent the birefringent component of the spindle.

The second component, the force-generating element of the spindle, is probably represented by some other molecular species. Rebhun (1967) has emphasized that the saltatory movements of particles in the spindle during division are very similar in character to movements that occur in other cells (e.g., cytoplasmic streaming in amebas, algae) and under other conditions (e.g., during interphase, in the presence of colchicine) where organized microtubules are absent. In cells in which cytoplasmic streaming, contraction, or movement of cytoplasmic particles occurs, studies with the electron microscope have often revealed the presence of numerous oriented microfilaments, 50–75 A in diameter, in locations suggesting their direct involvement in the movements (e.g., Wohlfarth-Bottermann, 1964a,b; McManus and Roth, 1965; Rhea, 1966; Wohlman and Allen, 1968, amebas and myxomycete plasmodia; Cloney, 1966, epidermal cells of ascidian embryos; Nagai and Rebhun, 1966, algal cells; Peters and Vaughn, 1967, nerve cells). Recently Bajer and Molè-Bajer (1969) have observed associations between spindle microtubules and bundles of microfilaments in *Haemanthus* endosperm cells. These observations, as well as the presence of microtubules in many cellular locations where such cytoplasmic movement probably does not occur (Porter, 1966; Gibbins *et al.*, 1969a), suggest that the spindle microtubules do not generate the mitotic forces but that they serve to orient the elements that do.

9. Proportionality between Pulling Force and Length of Kinetochore Fiber

The strength of the pull on a kinetochore increases with increasing distance from the pole to which the kinetochore is connected. This generally holds true, *on the average*, although other factors such as short-term local changes in the force-generating mechanism or differences in the numbers of kinetochore microtubules may override this "force-distance rule" for short periods of time or in special circumstances.

The equatorial position of the chromosome on the spindle at metaphase can be considered a position of stable equilibrium only if it is assumed that the chromosome is held there by antagonistic *pulling* forces that *increase* with increasing distance from the pole, or by antagonistic *pushing* forces that *decrease* with increasing distance from the pole (Rashevsky, 1941; Östergren, 1950). The latter alternative is ruled out for most cells by the evidence that kinetochores are *pulled* toward the pole throughout most or all of division (Chapter IV, Section B,3).

Something more can be said about the relation between the position of the kinetochore and the magnitude of the poleward force acting on it during prometaphase from the observations of Bauer *et al.* (1961) and of Dietz (1969) on crane fly spermatocytes. Univalents moving toward a pole in prometaphase come to rest at a distance of 1–4 μ away from the edge of the centrosphere (see Fig. 1 in Dietz, 1969), indicating that the net force acting on kinetochores in that position is zero. Bivalents and multivalents in which all kinetochores are temporarily oriented toward only one pole behave in the same way. The positions of trivalents at metaphase in spermatocytes of *T. oleracea* (in which the stretching of chromosomes and the poleward displacement of trivalents is maintained until anaphase) provide further information on the force-distance rule. The observations of Bauer *et al.* (1961) on four V- or Q-shaped trivalents that remained stably oriented for at least a few minutes are presented in Table VI. At equilibrium the forces on kinetochores b and c are balanced by the opposing force on kinetochore a (see diagram accompanying Table VI). The data are in reasonable

TABLE VI

Position of Kinetochores in Trivalents at Metaphase I in Spermatocytes of *Tipula oleracea*[a]

Figure in Bauer *et al.* (1961)	Equator-to-pole distance (μ)	Kinetochore-to-pole distance (μ)[b]			Relative force acting on kinetochore[b, c]		
		a	b	c	(a)	(b + c)	Difference
5	12.2	14.2	4.7	7.0	12.2	11.7	+0.5
6	12.7	16.4	3.9	5.9	12.7	9.8	+2.9
7	12.6	15.8	4.6	5.9	12.6	10.5	+2.1
9	10.3	12.3	2.8	5.3	10.3	8.1	+2.2

[a] Data from Bauer *et al.*, 1961.

[b] The positions of kinetochores a, b, and c are indicated in the following diagram:

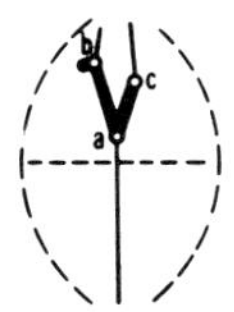

[c] Arbitrary units. The assumption is made that the force is directly proportional to distance from pole but that the distance used in the proportionality cannot exceed the equator-to-pole distance.

agreement with the idea that the poleward force acting on the individual kinetochore is *directly proportional* to its distance from the pole, if it is assumed that the distance used in the proportionality cannot exceed the length of the half-spindle (for the justification of this assumption, see Chapter IV, Sections B,10 and C,3 on the polarity of kinetochore fibers). The remaining discrepancies between the figures under column (a) and those under column (b + c) can be accounted for if it is remembered that the trivalent itself, in its off-equatorial position, is subject to a nonspecific poleward force independent of the forces acting on kinetochores, similar to the force that moves acentric chromosome fragments poleward. Some justification for this explanation can be found in the fact that agreement is best for the trivalent that lies closest to the equator (first row, Table VI).

These observations suggest that the force acting on a kinetochore is directly proportional to the length of the kinetochore fiber in the half-spindle, other factors being equal.

Does the proportionality between the force acting on the kinetochore and the length of the kinetochore fiber hold also for anaphase? If it does, and if it is assumed that the rate of chromosome movement in anaphase is linearly related to the force acting on its kinetochore, then the chromosomes might be expected to approach the poles with decreasing velocity. In published descriptions of anaphase movement in living cells, in which account has been taken of the contribution made by spindle elongation, this is sometimes the case. In the chick tissue culture cells studied by Hughes and Swann (1948), the rate of movement of chromosomes to the poles appeared to decrease linearly with the decreasing length of kinetochore fibers. Ris (1949) observed a decreasing rate of kinetochore separation in spermatocytes of the grasshopper *Chorotophaga viridifasciata* in which spindle elonation had been inhibited with chloral hydrate.

The linear relation between kinetochore-to-pole distance and rate of poleward movement during the course of anaphase may not always hold. In the spermatocytes of other grasshoppers, the chromosomes appear to approach the poles at a *constant* rate (Jacquez and Biesele, 1954; Makino and Nakanishi, 1955; Izutsu, 1960). The same is true for the spermatocytes of the beetle *Popilius disjunctus* (observations of H. S. Roberts, given in Fig. 63C of Mazia, 1961). It is possible that in these studies the locations of the poles of the spindle were misjudged; Forer (1966) found that in spermatocytes of the crane fly *N. suturalis* the spindle poles as seen in the polarizing microscope were often farther apart than they appeared to be in the same cell viewed in the phase-contrast microscope. Bajer (1961) has suggested that in some cells the kinetochore fiber may extend only part of the way to the pole, as appears to be the case from his observations on *Haemanthus* endosperm cells with the polarizing microscope. This means that at least some of the poleward movement at anaphase could take place with constant kinetochore fiber length. Clearly, anaphase movement in these other cells ought to be reinvestigated by means of polarizing microscopy. It is also possible that other factors, such as changes in numbers and orientation of kinetochore microtubules during the course of anaphase (see Bajer, 1968a,b), might influence the rate of anaphase movement in these cells. In some cases the force being applied to the kinetochore may not always be faithfully reflected in the velocity of the chromo-

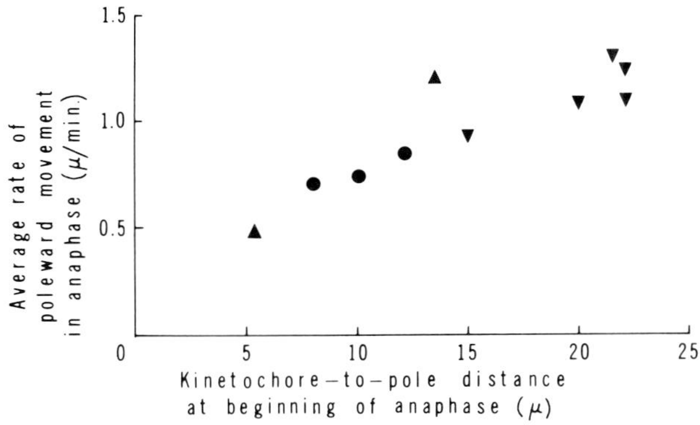

FIG. 18. Relation between anaphase chromosome velocity and distance from the pole. ▼, *Chorotophaga viridifasciata* spermatocytes; data from Nicklas (1963), fig. 8; ●, *T. oleracea* spermatocytes; data from Bauer *et al.* (1961), table 5; ▲, *T. oleracea* spermatocytes, dyads segregating from trivalents; data from Bauer *et al.* (1961), p. 154.

some, as for example under conditions of changing viscosity in the spindle.

Studies on the variation in rate of chromosome movement with distance from the pole *during the same interval of anaphase* avoid some of these difficulties. The data of Bauer *et al.* (1961) and of Nicklas (1963) provide independent evidence for a linear relation between the rate of anaphase movement and the distance from the pole. The dyads segregating from a bivalent or trivalent that lies off the equator at the onset of anaphase approach their respective poles at different rates. The kinetochores of a metaphase bivalent with a proximal chiasma lie closer to the equator (and farther from the poles) than do the kinetochores of a bivalent with a single terminal chiasma. Dyads segregating from the former approach the poles faster than those segregating from bivalents with only terminal chiasmata (see Fig. 18). A similar relation between the rate of anaphase movement and distance from the pole is obtained when dyads disjoining from bivalents close to the interpolar axis are compared with those disjoining from bivalents far from the interpolar axis (Nicklas, 1965).

In short, most of the evidence is in reasonable agreement with the idea that the force on the kinetochore is directly proportional to the length of the kinetochore fiber.

10. Polarized Structure of Microtubules

Microtubules are polarized. Although they are not always found where cytoplasmic motility occurs (see references in Chapter IV, Section B,8), it is a striking fact that wherever microtubules *are* associated with motility the net movement occurs parallel to the microtubules *in one direction only*; or, in those cases in which movements in both directions occur, electron micrographs have usually revealed two distinct sets of microtubules.

One-way movement occurs in asters and half-spindles (Chapter IV, Section B,6). It is found during pronuclear fusion in the sea urchin egg, in which the female pronucleus migrates toward the male pronucleus along the rays of the sperm-monaster (Chambers, 1917; E. B. Wilson, 1925, p. 394; see also Longo and Anderson, 1968). It is also seen in nuclear migration after cell fusion (Holmes and Choppin, 1968), in nerve axons (Schmitt, 1968), and possibly in the transport of vesicles from the Golgi apparatus (Lentz, 1967). During the "contraction" of melanocytes, melanin granules move from peripheral positions in the arms along channels defined by microtubules and aggregate in the central region of the cell (Bikle *et al.*, 1966); there is some evidence that the movement of the melanin granules in the opposite direction during "expansion" of the melanocyte represents a purely passive dispersal not requiring the presence of microtubules (Malawista *et al.*, 1966; Marsland and Meisner, 1967; Wikswo and Novales, 1969).

There is movement in opposite directions away from the equatorial region toward the two daughter nuclei in telophase (e.g., Stich, 1954; Beermann, 1959; Bajer and Molè-Bajer, 1963; Molè-Bajer, 1965; Byers and Abramson, 1968). Electron micrographs show that two sets of microtubules extend poleward from the equatorial region; very few individual microtubules extend without interruption from one half-spindle across the phragmoplast or midbody into the other half-spindle (Pawaletz, 1967; Byers and Abramson, 1968; Hepler and Jackson, 1968). Double systems of microtubules are also found in the tentacles of the suctorian *Tokophrya* (Rudzinska, 1965) and in the axopods of the heliozoan *Actinosphaerium* (Tilney and Porter, 1965), in which bidirectional movement occurs. The same may be true for the marine alga *Caulerpa prolifera* (Sabnis and Jacobs, 1967).

Whether the microtubules produce the motive force or, as seems more likely, induce a specific orientation in some other force-generating element, they clearly induce a polarization of movement that betokens a polarity in the structure of the microtubules themselves. Polarity in helical structures is in fact a necessary consequence of their being constructed from identical asymmetric subunits all forming the same bonding pattern with neighboring subunits [examples may be found in the polarized helical structures formed by tobacco mosaic virus protein (Caspar and Klug, 1962), sickle-cell hemoglobin (Murayama, 1966), and actin (Hanson and Lowy, 1963; Huxley, 1963)]. The observations presented in the preceding paragraph demonstrate that the structural polarity expected in microtubules is relevant to their function in motility.

The microtubules of kinetochore fibers are no exception to these generalizations, and some recent speculations have included kinetochore fiber polarity as a basic feature of the mechanism responsible for chromosome movements (Östergren *et al.*, 1960; Subirana, 1968; McIntosh *et al.*, 1969). Evidence for the functional polarity of kinetochore fibers can be found in the equilibrium position of trivalents in spermatocytes of *Tipula oleracea* and in the behavior of the univalent X chromosome in spermatocytes of the grasshopper *Melanoplus differentialis*. In the former case, the trivalent lies closer to one pole than could be accounted for on the assumption that the pulling force is strictly proportional to the length of the kinetochore fiber. The data on kinetochore position is better explained by assuming that the poleward forces effectively act only on the portion of the kinetochore fiber in the half-spindle containing the pole toward which the kinetochore is oriented (Table VI, Chapter IV, Section B,9). A similar situation arises just after reorientation of the grasshopper X chromosome at one of the poles. If this assumption is correct, then it might be expected that during the first half of a pole-to-pole trip, when the chromosome is moving to the equator, its velocity would be relatively high, decreasing linearly only after the chromosome crosses the equator. The behavior of the X chromosome in *Melanoplus* often appears to follow this pattern (see Nicklas, 1963, Fig. 3; 1965, Fig. 7), although more extensive data would be required to distinguish this model unequivocally from one in which forces acting on the whole length of the kinetochore fiber contribute to the poleward pull on the kinetochore.

11. Anaphase Chromosome Movement without Spindle Poles

Organized spindle poles are not necessary for anaphase chromosome movement.

All degrees of compactness of the poles of the spindle, from highly localized regions to broad diffuse areas as wide or wider than the metaphase plate itself, are compatible with anaphase movement of the chromosomes. Such variation is found even within a single cell type: in different pollen mother cells of hybrid or mutant plants (Darlington and Thomas, 1937; Clark, 1940), as a natural variation in the early divisions after fertilization in *Pinus strobus* (Ferguson, 1901b), as a consequence of various degrees of flattening in endosperm cells (e.g., Inoué and Bajer, 1961), and in tipulid spermatocytes (Dietz, 1959, 1963; and Fig. 29 in Went, 1966a).

Anaphase separation can also occur in monopolar figures, where one chromatid moves toward the pole and the other appears to move away (Bělař and Huth, 1933). A striking example of the degree to which the whole spindle can be disorganized without preventing anaphase separation of chromosomes is found in the "apolar" mitosis of *Haemanthus* endosperm cells treated with chloral hydrate (Molè-Bajer, 1967, 1969).

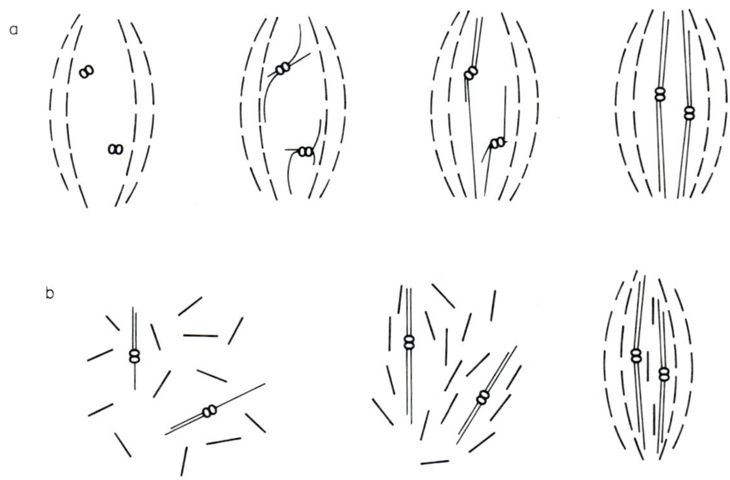

FIG. 19. Two mechanisms of chromosome orientation in which kinetochore fibers are formed by the kinetochores. Two chromosomes are shown as if transected through the kinetochore region. (a) Top row, four successive stages of prometaphase, in which the orientation of kinetochore fibers is determined by already oriented continuous fibers and by poleward pulling forces transmitted to the kinetochores by the kinetochore fibers. (b) Bottom row, three stages of prometaphase, in which the orientation of kinetochore fibers, and secondarily of continuous fibers, is determined by the kinetochores themselves. Mutual alignment of the kinetochores of different chromosomes is achieved by fusion of the individual chromosomal spindles.

No well-defined poles are recognizable and continuous microtubules are largely absent. No metaphase plate is formed and the orientation of the kinetochores of one chromosome often shows little relation to the orientation of nearby chromosomes. Nevertheless, anaphase occurs and sister kinetochores move apart in opposite directions, with kinetochore microtubules leading the way. The velocity of chromosome movement is one-fourth to one-half that of normal anaphase.

C. A MODEL OF CHROMOSOME ORIENTATION AND MOVEMENT

Most of the observations in the previous sections can be incorporated into a general model of chromosome behavior, illustrated in Figs. 19 and 20. The main features of the model are as follows.

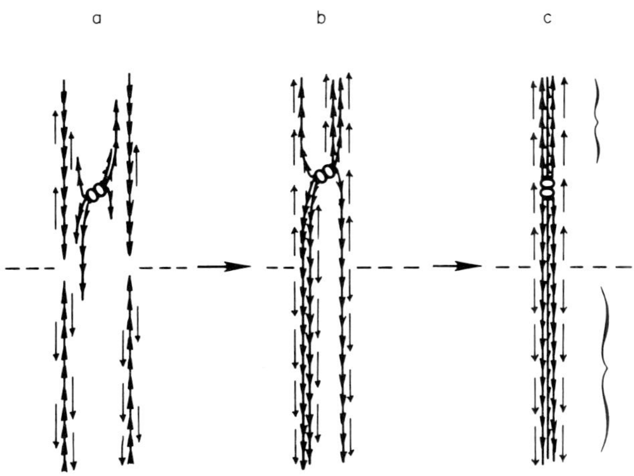

FIG. 20. A model of chromosome orientation and congression involving the interaction of force-generating elements with polarized microtubules. One chromosome is shown in three successive stages of prometaphase. Polarity of microtubules is indicated by arrowheads, force-generating elements by thin arrows. The dashed horizontal line represents the equator of the spindle. (a) Early prometaphase; force-generating elements are oriented by polarized continuous microtubules. (b) Mid-prometaphase; oriented force-generating elements act on polarized kinetochore microtubules to align sister kinetochores in the spindle axis. (c) Late prometaphase; bracketed regions show the portions of microtubules interacting effectively with the force-generating elements; the pair of sister kinetochores moves toward the equator of the spindle.

1. Microtubule Polymerization and Orientation

In some cells continuous fibers are assembled from microtubular subunits around local centers often associated with centrioles or other polar structures. These centers play only an indirect role by providing local conditions favorable for subunit polymerization. In other cells polymerization occurs "spontaneously" at many different sites, giving rise to areas of randomly oriented microtubules. In either case, through lateral associations of individual microtubules, bundles of microtubules gradually line up in parallel to form the continuous fibers of the developing spindle (see Chapter II, Section C,6).

2. Formation of Microtubules at Kinetochores

New subunits are added to the elongating microtubules of the kinetochore fiber primarily at the kinetochore. In cells normally without definite polar centers for microtubule polymerization, or in experimental situations in which such centers are lacking, the kinetochore microtubules extend outward approximately perpendicular to the kinetochore surface and influence the orientation of other microtubules in the vicinity. In cells with active centers the orientation of the newly forming kinetochore microtubules is largely determined by the continuous microtubules already oriented in parallel between the two poles (Fig. 19). These two modes of orientation of the kinetochore microtubules are not mutually exclusive; in some cells both may occur in different regions of the developing spindle, depending on the degree to which continuous fibers are already oriented locally.

These mechanisms of kinetochore fiber orientation were suggested by Pease's study of *Tradescantia* pollen mother cells fixed a few minutes after the termination of hydrostatic pressure treatment: "Invariably the base of the traction fiber extended directly away from the kinetochore, and it did not bend towards a pole until it became associated with continuous fibers. The bend would then be towards the pole less than ninety degrees away from the initial growth direction even if this happened to be the 'wrong' pole. It thus looked as though the growth direction was unimpeded until the traction fiber became associated with continuous fibers, and then the further extension of the traction fiber followed the path of least resistance in the pattern expressed by the continuous fibers" (Pease, 1946). The only modification of this view required by more recent observations is that in early division stages individual microtubules from a single kinetochore may extend toward opposite poles (Luykx, 1965b; Bajer and Molè-Bajer, 1969).

3. The Role of Kinetochore Microtubules

The kinetochore microtubules are purely structural elements; they transmit, rather than generate, the poleward force that acts on the kinetochore.

Kinetochore microtubules are polarized in such a way that the force-generating elements interact effectively with the kinetochore microtubules only when both are appropriately oriented; if their mutual orientation is reversed, no force is applied to the microtubules. The force acting on the kinetochore depends on the microtubular surface area available for effective interaction with the force-generating component. It therefore depends on the number of kinetochore microtubules, their length, and their orientation relative to that of the force-generating elements (Chapter IV, Sections B,9 and 10).

4. Chromosome Orientation and Congression

Because of variations in the numbers of kinetochore microtubules extending toward the poles and in the force applied locally to the kinetochore microtubules, a maloriented chromosome is subject to fluctuations in poleward forces leading ultimately, through a series of detachments and re-formations of kinetochore microtubules, to a relatively stable orientation of the chromosome, the orientation in which the two sister kinetochores (or homologous kinetochore pairs, in meiosis) are aligned in the spindle axis (Fig. 20). Discussion of the details of this mechanism of orientation can be found in the papers of Östergren (1951), Luykx (1965b), and Nicklas and Koch (1969).

At the same time, because the kinetochore microtubules are polarized (Chapter IV, Section B,10) and can interact with the force-generating elements only when the two are appropriately oriented with respect to one another, the chromosome lying off the equator is subject to a net force that shifts it toward the equator (Fig. 20).

In some cells the parallel orientation and equal length of kinetochore microtubules extending toward the two poles may be determined not by the poleward forces but more directly by the kinetochores themselves. The parallel alignment of the microtubules associated with kinetochores of different chromosomes and the equatorial positioning of all the chromosomes in such cells may be achieved by the fusion of the individual chromosomal spindles (Fig. 19b; see also Chapter II, Section C,6 and Chapter III, Section B,5).

5. The Orientation of the Force-Generating Elements

The force-generating component of the mitotic apparatus is closely associated with, and oriented by, all spindle microtubules other than those organized by the kinetochores.

It might at first be imagined that the force-generating elements are *not* preferentially directed toward one pole or the other in any region of the spindle, and that polarization of the force applied to the kinetochore derives solely from the interaction of structurally polarized kinetochore microtubules with those force-generating elements that do happen to be appropriately oriented. Such a concept, however, is incompatible with the observation that acentric chromosome fragments and various kinds of cytoplasmic particles are moved poleward at the *same rate* as anaphase chromosomes. It must be presumed therefore that the force-generating elements as well as the microtubules are polarized. These elements must become oriented parallel to the interpolar axis in the spindle and radial to the poles in asters in such a way that the force is directed toward the nearest pole. They must be close enough to kinetochore microtubules to interact with them. Finally, they must be more firmly anchored in the spindle than the kinetochore microtubules in order for the force to be effectively translated into kinetochore fiber movement.

In their locations and orientations in the developing spindle, the microtubules of the asters and continuous fibers show a number of similarities to the force-generating elements. If they are formed first in the polar regions of the developing spindle, the continuous microtubules may be polarized in a sense opposite to that of the kinetochore microtubules. Continuous microtubules from opposite poles may interact or mutually terminalize each other in the equatorial region to provide a relatively rigid framework from which the poleward-oriented forces can act on kinetochore microtubules (Fig. 20a). The close association of continuous and kinetochore microtubules seen in many cell types (see Chapter IV, Section A,2) may be functionally significant in this connection. Nevertheless, the continuous microtubules are probably not themselves the force-generating component of the spindle, because anaphase movements can take place in the absence of continuous fibers and because in many cells similar (but nonmitotic) cytoplasmic movements occur without any microtubules at all (Chapter IV, Sections B,8 and 11).

Where continuous microtubules are absent, the force-generating elements will be oriented at random. Only a small fraction of these elements would then be able to interact effectively with polarized kinetochore microtubules, and it would be predicted that under such circumstances anaphase chromosomes would move at a fraction of their normal speed. This is precisely what is seen in *Haemanthus* endosperm cells treated with chloral hydrate (Molè-Bajer, 1967, 1969). These considerations argue against the proposal of Östergren *et al.* (1960) that the *kinetochore* fibers themselves orient the force-generating elements. It can be imagined that kinetochore microtubules are unable to orient the force-generating component because of their association with some masking material, perhaps a product of synthetic activity by the kinetochore itself. The difference between kinetochore microtubules and continuous microtubules, with respect to their sensitivity to agents such as colchicine, chloral hydrate, and low temperature, provides some support for this idea (Chapter IV, Section A,2).

6. Variation in Kinetochore Activity

Since acentric chromosome fragments and cytoplasmic particles move poleward during *metaphase*—at a time when the poleward force acting on the kinetochore is supposed to be at a minimum (Chapter IV, Section B,4)—it must be assumed that at metaphase the poleward force in some way becomes disengaged from the kinetochore. It could be that the force-generating component becomes disengaged from kinetochore microtubules as metaphase approaches, but the observation of Forer (1965, 1966) that birefringent kinetochore fibers appear to be moving poleward even during metaphase makes this seem unlikely. More attractive is the possibility that there is a change in the association between the kinetochore microtubules and the kinetochore. It is proposed that the rate of microtubule assembly at the kinetochore varies in a regular way during the course of division and that this variation affects the net poleward force acting on the kinetochore.

During prometaphase the rate of microtubule assembly at the kinetochore is low. Microtubules are moved poleward faster than they are assembled at the kinetochores, and the kinetochore regions of appropriately oriented chromosomes are stretched poleward. As metaphase approaches, the rate of microtubule assembly at the kinetochore increases until at metaphase it approximates the rate at which the microtubules are moved poleward; the poleward force acting on the kinetochore is zero.

The kinetochore might actively determine the rate of assembly of kinetochore microtubules. This could account for the cases in which chromosomes move away from the pole (or poles) to which they are attached (Metz *et al.*, 1926; Scott, 1936; Schrader, 1946a,b). In these cases the rate of microtubule assembly at the kinetochore exceeds the rate at which the microtubules are moved poleward. In the more usual case the rate of microtubule assembly at the kinetochore during metaphase would have to be closely regulated in order to match the rate at which the microtubules are moved poleward. Another possibility is that the maximum rate of assembly is determined primarily by the rate of poleward movement of the microtubules, the kinetochore acting merely as a kind of valve controlling the influx of subunits into the microtubules.

Around the time sister chromatids separate, assembly of microtubules at the kinetochore ceases or is substantially reduced, and the kinetochore is subject to the full poleward force acting on the kinetochore microtubules. Variations in the time at which this kinetochore activity ceases, relative to the time of chromatid separation, determine whether there is an early anaphase stretch or whether chromatids simply separate passively a few minutes before they move poleward (Chapter IV, Section B,4).

The firm attachment of the chromosome to the spindle (Chapter IV, Section B,1) is compatible with the assembly of microtubules at the kinetochore, if the site of attachment is different from the site of assembly. For example, in the kinetochore of animal cells (Chapter III, Section C,6,*a*) the attachment and assembly sites might be the outer and inner dense layers, respectively.

The increase in compactness and electron density or osmiophilia seen in the kinetochores during prometaphase of some organisms (Chapter III, Section C,6,*a*) may be related to the increased rate of microtubule assembly at the kinetochores. No structural changes in the kinetochore at the onset of anaphase have been reported. As Forer (1966) noted, however, the distance between the kinetochore and an area of reduced birefringence on the kinetochore fiber produced by UV microbeam irradiation increased not at all, or much more slowly, in anaphase than in metaphase, implying a much reduced kinetochore activity after anaphase begins. Nicklas and Staehly (1967) found that the attachment of chromosomes to kinetochore fibers appeared to be firmer during anaphase than during prometaphase or metaphase.

D. DIETZ'S MODEL

The major weakness of the model presented above is that it does not satisfactorily account for certain observations of Dietz (1958, 1969; see also Bauer *et al.*, 1961), namely, that (1) the movement of univalent chromosomes in prometaphase may be up to three times faster than dyad movement in anaphase; and (2) trivalents often take up a position *on the equator* of the spindle at metaphase. If both prometaphase and anaphase chromosome movements are a result of the same poleward force, it is not clear why the velocity of prometaphase chromosomes can exceed that of anaphase chromosomes, although changes in the kinetochore or in the number and arrangement of kinetochore microtubules might account for the difference. Second, a trivalent displaced toward one pole during prometaphase would not automatically return to the equator as the pulling forces acting on the kinetochores decrease toward metaphase. Östergren (1961) has speculated that the prometaphase history of the trivalent might induce functional differences among its kinetochores such that the force acting on the kinetochore pair that faces one pole alone might temporarily overcome the combined forces acting on the opposing kinetochore pairs, but there is no independent evidence to support this idea.

These observations are more readily accounted for in a detailed hypothesis recently set forth by Dietz (1969). The hypothesis describes the course of division in the following way. As the cell enters division, spindle proteins "crystallize" out in the form of microtubules, arising first at the surface of the kinetochore, which somehow prescribes the formation of secondary bonds between spindle microtubule subunits. Microtubules represent thermodynamically stable structures, with subunits arranged so as to provide a maximum number of intermolecular bonds. At their polar ends (furthest from the kinetochore), where there are fewer intermolecular bonds, the microtubules are anchored in an unoriented gel of microtubule subunits. Microtubules grow longer by the addition of more subunits throughout their length. As the number of free subunits available for incorporation into microtubules becomes limiting and the rate of microtubule elongation decreases, the number of intermolecular bonds continues to increase primarily by the formation of greater numbers of microtubules in both chromosomal and continuous fibers. Since there is a continuous exchange of subunits between existing microtubules, unoriented gel, and a pool of free molecules, the formation of new microtubules under conditions where the number of free subunits is limiting involves the recruitment of some subunits from already existing kinetochore microtubules. The subunits remaining in such partially disassembled microtubules spontaneously rearrange themselves to give shorter microtubules, again with a maximum number of intersubunit bonds. The shortening of the kinetochore microtubules, increasing in number and anchored in the polar regions, exerts a gradually increasing pull on the kinetochores, resulting in the prometaphase stretch. As the cell progresses toward metaphase, the number of kinetochore microtubules reaches the maximum that can be accommodated on the kinetochore surface, new kinetochore microtubules are no longer formed, the amount of spindle material incorporated into microtubules reaches a maximum, and the kinetochore fiber no longer exerts a pull on the kinetochore. Since each kinetochore induces the formation of approximately equal numbers of intermolecular bonds, kinetochore fibers grow to approximately equal lengths and thereby ordain the equatorial position of the chromosome (bivalent or trivalent) at metaphase. At the onset of anaphase sister chromatids part and the subunits of the microtubules are slowly solubilized, those left behind in the shortening kinetochore microtubules continually rearranging themselves to maintain the thermodynamically stable microtubular structure. The rate of shortening of the kinetochore microtubules in anaphase is limited by the rate of solubilization of microtubular subunits; hence, the poleward movement of dyads in anaphase may be slower than that of univalents in prometaphase, where the rate is limited primarily by the rate of subunit reassembly in partially formed microtubules.

The evidence that the rate of anaphase chromosome movement is proportional to the length of the kinetochore fiber (Chapter IV, Section B,9) can be accounted for by assuming that the rate of solubilization of subunits per unit length of microtubule is constant throughout most of anaphase. A well-defined spindle pole would not be necessary for anaphase movement (Chapter IV, Section B,11) because kinetochore fibers would be anchored in unoriented gel at their polar ends independently of one another.

There are two main objections to this hypothesis. The first is that it does not readily explain the observations indicating that microtubules do not contract along their entire length during anaphase shortening (Chapter IV, Section B,7). If the hypothesis is modified so that anaphase shortening is accomplished by the disassembly of kinetochore microtubules only at their polar ends, then it would predict that the rate of anaphase movement would be independent of the length of the kinetochore fiber, in contradiction to the observations noted in Chapter IV, Section B,9. The second objection to this hypothesis is that it ignores the fact that a wide variety of cytoplasmic particles are also moved poleward (Chapter IV, Section B,6) during metaphase as well as anaphase. It is of course possible that this poleward movement is unrelated to the movement of chromosomes, but since the rates are approximately the same this seems unlikely.

As an aid to designing experiments to distinguish between these two hypotheses, Table VII summarizes their major points of difference.

TABLE VII

Two Hypotheses of Chromosome Movement During Mitosis

Feature of mitosis	"Two-component" hypothesis (Chapter IV, Section C)	"One-component" hypothesis (Chapter IV, Section D)
Origin of poleward forces acting on kinetochores	Force-generating elements, ordinarily oriented toward poles, acting on kinetochore microtubules	Contraction of kinetochore microtubules themselves
Basis for poleward movement of chromosomes	Polarized kinetochore microtubules extend into regions containing force-generating elements	Kinetochore microtubules, not necessarily polarized, are anchored at poles
Cause of equatorial position of chromosomes at metaphase	Force acting on sister kinetochores is proportional to kinetochore fiber length	Fibers of equal length grow from sister kinetochores
Reason for minimum force acting on kinetochores at metaphase	Increased rate of microtubule assembly at kinetochore compared to prometaphase	Decreased contraction of kinetochore microtubules
Change at onset of anaphase	"Tightening" of connection between microtubules and kinetochore	Beginning of solubilization of kinetochore microtubule subunits

Both schemes deal mainly with the problems of chromosome orientation, metaphase equilibrium, and chromosome movement to the poles. In many cells the elongation of the spindle also makes an important contribution to anaphase chromosome separation, but spindle elongation has not been studied in any detail and is not specifically accounted for in either of the two hypotheses presented above. It is clear that continuous microtubules are present in every spindle; and the fact that drugs such as colchicine and chloral hydrate preferentially disrupt continuous microtubules and simultaneously prevent spindle elongation (Chapter IV, Section A,2) implicates these microtubules in the elongation mechanism. Bélař's (1929) studies on the effects of hypertonic solutions on dividing grasshopper spermatocytes suggest that changes in the state of hydration of spindle proteins are involved, a possibility consistent with Dietz's hypothesis. It would be interesting to reinvestigate this finding by means of polarizing and electron microscopy. The sliding of microtubules past one another that is conjectured to be the mechanism of cell elongation after contraction in *Stentor* (Bannister and Tatchell, 1968) might also serve as a model of spindle elongation (see also McIntosh *et al.*, 1969).

E. NONRANDOM CHROMOSOME SEGREGATION

In general, individual chromosomes behave independently unless physically connected. The random orientation of maternal and paternal homologs to one or the other of two poles in meiosis (accounting for Mendel's second law), the regular segregation of paired homologs to opposite poles, and the random assortment of homologous chromosomes to the two poles when they are not paired (see Weber, 1969), all are in accord with this assertion. Sometimes the behavior of one chromosome influences that of others to a slight extent. For example, in *Haemanthus* endosperm, normal anaphase chromosomes close to daughter kinetochores retarded by a chromosome bridge may be slowed down themselves (Bajer, 1964); but such observations are attributable to divergence and intermingling of microtubules from adjacent kinetochores (Bajer, 1968a,b).

In some cells, however, daughter chromosomes may be segregated to daughter cells in a decidedly nonrandom fashion.

1. Mitosis

Lark *et al.* (1966) found, in radioautographic studies of primary cultures of mouse embryo fibroblasts, that tritium-labeled thymidine incorporated into nuclear material (presumably chromosomal DNA) is distributed almost exclusively to only one of the two poles during the second division after exposure of the cells to the label. A similar pattern of distribution has been observed in Chinese hamster cells (Lark *et al.*, 1966), in root tips of a number of strains of wheat (Lark, 1967, 1969), and in germinating conidia of the fungus *Aspergillus nidulans* (Rosenberger and Kessel, 1968). With semiconservative DNA replication and a semiconservative mode of distribution of chromatid subunits (Taylor, 1963), these observations mean that at metaphase the chromosomes are arranged in such a way that the chromatids containing the DNA synthesized two generations earlier are all oriented toward the same pole. However, in mouse KB cells and HeLa cells (Lark *et al.*, 1966), primary cultures of human blood cells (Cuevas-Sosa, 1968), spermatogonia of the newt *Triturus vulgaris* (Callan and Taylor, 1968), and root tips of certain other strains of wheat (Lark, 1967, 1969), similar experiments have suggested a *random* distribution of daughter chromosomes to daughter cells. [In roots of *Vicia faba,* Lark (1967) found evidence for nonrandom chromatid distribution, but Heddle *et al.* (1967) concluded from their studies that chromatid distribution in the same material is random. The reason for this discrepancy is not known.] The phenomenon of nonrandom chromatid distribution in mitosis is thus not universal, but its common occurrence is firmly established.

These results have revived interest in the hypothesis that the chromosome, the spindle pole, and the kinetochore fiber that connects the two represent a unit that persists throughout the cell cycle and is duplicated between divisons (Lettré and Lettré, 1958, 1959, 1960; see Goldstein, 1967). The separation of a newly formed pole from the old one would then automatically lead to the orientation of one set of chromatids toward one particular pole. This idea would not be applicable, however, to cells in which definite pole determinants are absent (Chapter II, Sections A,3 and C,5). Furthermore, it can be discounted for cells in general because it is not easily reconciled with the lability of kinetochore microtubules and their apparent absence between divisions, with the random orientation of maternal and paternal homologs in meiosis, or with the reorientation of chromosomes brought about by micro-

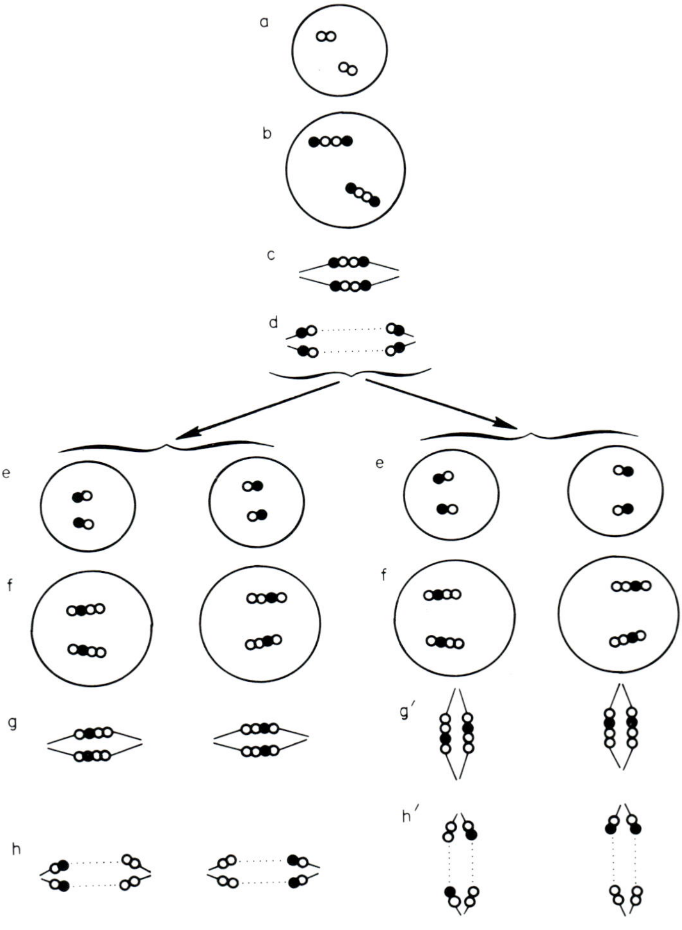

FIG. 21. A possible mechanism of nonrandom distribution of chromatids at mitosis. Two chromosomes are shown as if cross-sectioned through their kinetochore regions. Each small circle represents the cross-section of a chromatid subunit. The chromatid or unduplicated chromosome consists of two subunits. Open circles, unlabeled subunits; filled circles, labeled subunits. (a) Before chromosome duplication; (b) after duplication of chromosomes in labeled medium; (c) metaphase; (d) anaphase; (e) daughter nuclei; (f) second duplication of chromosomes, in unlabeled medium; (g,g′) metaphase; (h,h′) anaphase. (g,h) When the spindle axis is parallel to the axis of the spindle in the previous division and the kinetochore regions of the

manipulation or observed in the univalents and trivalents of untreated meiotic cells. The formally equivalent scheme in which there is a permanent functional difference between the two spindle poles and a complementary difference between the two kinetochores of each chromosome can be ruled out for some of the same reasons; this suggestion is also incompatible with the metaphase distribution of chromosomes observed in multipolar spindles. The nonrandom distribution of daughter chromosomes is better explained on other grounds.

In an important radioautographic study of cultured cells of the marsupial *Potorous tridactylis*, Walen (1965) demonstrated that during chromosome duplication the daughter chromatids are formed in such a way that newly synthesized DNA is usually located in the outer subunits of a pair of chromatids (see diagram, Fig. 21). Similar observations have been made on human cells (Schwarzacher and Schnedl, 1966; Herreros and Gianelli, 1967). These findings imply that the newly synthesized DNA is associated with the poleward half of the kinetochore of each anaphase chromosome. If the orientation of the kinetochore regions of the chromosomes does not change much during interphase, and if the spindle in the next division lies in the same axis as the spindle of the preceding division, then the orientation of kinetochores will tend to be preserved during the establishment of the metaphase plate, and at anaphase labeled daughter chromosomes will segregate from unlabeled ones (Fig. 21).

This idea is made plausible by the fact that in many cells the relative positions of whole chromosomes, or heterochromatic segments of chromosomes, does not change from telophase to the next prophase (E. B. Wilson, 1925, pp. 890–903; Barber, 1940; Beerman, 1959, 1966; Comings, 1968; Wagenaar, 1969). Carlson (1956) noted that the distal ends of chromosomes in prophase nuclei of grasshopper neuroblasts are more easily displaced with a microneedle than are the proximal ends, indicating that the kinetochores are held in a fixed position in the nucleus. Furthermore, in at least some of the cases in which nonrandom distribution of chromosomes occurs, it is known that the spindles of successive divisions are oriented in the same axis—e.g., in

---

chromosomes do not change their orientation during interphase, labeled chromatids separate from unlabeled chromatids. (g′,h′) When spindle axis is perpendicular to that of the previous division, kinetochores must reorient and chromatids segregate randomly.

root tips and in germinating conidia. *Random* chromatid distribution would be expected wherever (1) newly synthesized chromatid subunits are distributed to daughter chromosomes in a more random fashion than that described by Walen for *Potorous* cells; (2) kinetochore orientation is not preserved during interphase; or (3) the axis of the spindle is oriented at right angles to the spindle axis of the preceding division (Fig. 21).

Thus, the nonrandom segregation of anaphase chromosomes in mitosis requires the assumption neither of permanent connections between kinetochores and poles, nor of complementary functional differences between poles and kinetochores.

Two other examples of nonrandom chromosome distribution in mitosis deserve comment. Teplitz *et al.* (1968) have documented a case of interspecific hybridization of mammalian somatic cells in which multipolar division of the fused cells led to the production of daughter cells many of which contained a complete *haploid* set of chromosomes from each of the parental cell types. Temporary incubation at low temperatures caused another segregation of haploid chromosome sets within one or two cell generations, resulting in the reappearance of diploid parental cell types. The possibility of extensive chromosome rearrangements and selective growth in one of the parental cell types was ruled out by karyotype analysis and growth rate studies on control parental cell cultures. In kidney epithelial cultures of *Microtus agrestis* (field mouse), Pera and Schwarzacher (1969) found a number of cells undergoing tripolar division. Microphotometric measurements of Feulgen DNA and observations on the distribution of the distinctive X chromosome suggested that each pole was receiving one or more complete haploid sets of chromosomes. All these observations could be explained by assuming that connections between a heritable pole determinant and each kinetochore of a haploid chromosome set are maintained intact through the processes of cell fusion and multipolar division. An explanation more in line with the observations presented in the preceding paragraphs, however, is that haploid chromosome sets are somehow maintained as separate groups in the interphase nucleus, as are diploid "subnuclei" in the macronucleus of ciliates (see Nanney, 1964). Such haploid sets might become associated with independent spindles (see Chapter II, Section C,4) and the poles of some spindles may occasionally fuse with those of others, so that daughter nuclei containing two or more complete haploid sets arise from multipolar divisions.

There have been some claims that nonrandom arrangements of chromosomes in diploid cells sometimes involve separation of haploid sets (e.g., Wilson and Cheng, 1949). Observations on *Bellevalia romana*, in which the chromosomes can be unequivocally distinguished from one another, do not support this conclusion (see Gläss, 1961, 1962), but in view of these recent startling results on mammalian cells in culture, the earlier claims deserve careful reinvestigation.

2. Meiosis

Other problems are raised by examples of nonrandom orientation and segregation of certain meiotic chromosomes in the absence of any apparent physical contact between them. In some cases the segregation of seemingly unpaired chromosomes to opposite poles during prometaphase or metaphase has been shown to be only the result of a premature disjunction of paired elements (Cooper, 1951; Smith, 1953; Jackson and Cheung, 1967). In liverworts (Lorbeer, 1934) and many neuropteran insects (Naville and de Beaumont, 1933; Hughes-Schrader, 1969), however, the X and Y chromosomes are not paired in the late prophase nucleus, yet are regularly found at opposite poles of the spindle at metaphase and anaphase. The conclusion that in these cases regular segregation is achieved without physical contact between the X and Y could be challenged on the grounds that pairing might occur only briefly during prometaphase or metaphase and could easily be missed in fixed preparations. This explanation clearly does not apply to the spermatocytes of a number of species of tipulid flies, however, in which the behavior of the sex chromosomes has been observed continuously in the living cell (Bauer *et al.*, 1961; Dietz, 1956, 1969). During prometaphase the X and Y chromosomes migrate poleward, remain near one pole, or reorient and move back toward the equator or the opposite pole, more or less independently of one another. By late metaphase the two sex chromosomes have taken up positions on the spindle equator where they remain while the autosomal dyads move toward the poles in anaphase. At late anaphase or early telophase, the X and Y move from their adjacent positions at the equator toward opposite poles, each one maintaining birefringent spindle fibers extending toward each pole all the while. Similar events may characterize the segregation of sex chromosomes during anaphase (synchronously with the separation of autosomal dyads) in the heteropteran insect *Dysdercus koenigii* (Ray-Chaudhuri and Manna, 1952) and the prometaphase segregation of the sex chromosomes in liverworts and in neuropteran insects.

It is clear that any two chromosomes, paired or unpaired, could be oriented and distributed in a specific, predetermined manner, given (1) a difference between the two poles, or between the two half-spindles; *and* (2) a specificity of interaction with the two different poles (or half-spindles) that would allow the two chromosomes to distinguish between them. If the two chromosomes regularly segregate to *opposite* poles, then this mechanism must also involve a difference between the two chromosomes.

As the observations discussed in the previous section show, it is not necessary to postulate functional asymmetries in the spindle poles or in sister kinetochores in order to account for nonrandom distribution of chromosomes in mitosis. In a number of cases of nonrandom segregation of unpaired chromosomes in meiosis, likewise, one need not assume permanent or regularly recurring functional differences between the two poles and between kinetochores. Nevertheless, such asymmetries are demonstrable in a few cases and may be more widespread than is generally realized. There are several examples of cells in which one pole or half-spindle is organized before the other (Hertwig, 1899; Ferguson, 1901a; Storch, 1924; Metz, 1938), and there is some evidence that in prometaphase of mitosis the two sister kinetochores of a chromosome may also develop asynchronously (Jokelainen, 1965). A regular asynchrony in the development of the two poles, and a concurrent asynchrony in the development of kinetochores, could result in the pole-specific orientation of chromosomes. There is no conclusive evidence for the operation of such a system in any cell, but some support for the idea can be found in the two successive meiotic divisions in *Sciara* spermatocytes and in certain cases of "meiotic drive" in *Drosophila* spermatocytes.

In *Sciara* (Metz *et al.*, 1926; Metz, 1933, 1938) the spermatocytes are radially disposed within a cyst. In the first meiotic division only a single pole is formed in each spermatocyte. All the chromosomes, unpaired since early prophase, are arranged with their kinetochores oriented toward the pole, yet maternal and paternal homologs respond differently to that pole. Maternal chromosomes remain oriented close to the pole, while paternal chromosomes back *away* from the pole in asynchronous fashion. These events are oriented in such a way that in all spermatocytes paternal chromosomes move toward the center of a cyst and are finally pinched off in small cytoplasmic buds. The spindle for the second meiotic division is bipolar, but in each secondary spermatocyte one pole—the one farther from the center of the cyst—appears to form earlier than the other. One of the chromosomes does not take part in the formation of the equatorial plate but always moves precociously (and undivided) to the earlier formed pole. As noted earlier (Chapter III, Section C,5), the characteristic behavior of this chromosome appears to be determined by a short heterochromatic chromosome segment adjacent to the kinetochore region.

Thus, in the first meiotic division two sets of chromosomes respond differently to a single pole, while in the second meiotic division a single chromosome responds preferentially to one of the two poles. There must be a functional difference between the two sets of chromosomes in the first meiotic division and a functional difference between the two poles in the second.

"Segregation distortion" in males of *D. melanogaster* is determined by a locus, SD, in or near the proximal heterochromatin of chromosome II (Sandler *et al.*, 1959). Markers adjacent to SD are found in more than one-half of the progeny of heterozygous males; in some lines the frequency of recovery of the SD-bearing chromosome is as high as 99%. The failure to recover the homologous chromosome (carrying the normal allele, SD^+) in the expected numbers is not a result of the preferential death of eggs fertilized by SD^+ sperm, for the percentage of eggs hatching in crosses showing segregation distortion is approximately the same as in crosses not showing it. Cytological studies of the first and second meiotic divisions in spermatocytes of heterozygous males reveal no abnormalities; furthermore, in males carrying a translocation between chromosome II and chromosome IV, the cytologically recognizable SD^+ chromosome occurs in about half of the second division cells, thus ruling out the possibility that it is regularly lost during development of the testis or during the first meiotic division (Peacock and Erickson, 1965). A possible explanation for the phenomenon is that the two daughter cells produced by the first meiotic division are normally different, one ultimately giving rise to two functional sperm cells, the other to two nonfunctional sperm cells. Segregation distortion would occur if the SD^+ chromosome were usually distributed to the daughter cell giving rise to nonfunctional sperm and its homolog (SD) to the other daughter cell. Such a mechanism implies a difference between the two poles of the first meiotic spindle, *and* a means of orienting the SD/SD^+ bivalent in a specific relation to the two poles. Peacock and Erickson (1965) supported this idea by showing that although four morphologically normal sperm are formed from each primary spermatocyte the number of progeny produced by females is only about one-half the number of inseminating sperm (in normal as well as in SD stocks). A functional difference between the two poles is also suggested by the finding that in males whose spermatocytes contain bacteria the bacteria are mostly located in the vicinity of one of the two poles (Peacock and Erickson, 1964; Yanders *et al.*, 1968; Erickson and Acton, 1969)—an observation reminiscent of the monopolar distribution of mitochondria in certain cells of *Chara* (Pickett-Heaps, 1967). A difference between the two homologous chromosomes at the SD locus appears to be necessary, for segregation distortion is not shown by homozygous males.

The support for this proposition is weakened, however, by several considerations. First, localization of bacteria primarily at one of the two poles does not necessarily imply a functional difference between the two poles; it could reflect only a previous nonrandom localization followed by elimination toward the nearest pole during spindle formation. Second, the ratio of inseminating sperm to progeny produced is influenced to a considerable degree by the genotype of the female and therefore may not always be a valid indication of events taking place during sperm formation in the male (Zimmering and Fowler, 1968; Denell and Judd, 1969). Third, segregation distortion in heterozygous males may be accompanied by a partial loss of fertility. Mated to females taken from isogenic stocks, males in whose progeny the normal (SD^+) chromosome is recovered only rarely, produce only about one-half as many offspring as do males in whose progeny the frequency of the SD^+ chromosome approaches the normal 50% (Hartl *et al.*, 1967). Homozygous males are almost completely sterile (Hartl, 1969). These observations suggest that the mechanism of segregation distortion involves the inactivation of SD^+-bearing sperm rather than the preferential segregation of the SD^+ homolog into normally nonfunctional sperm.

In another example of meiotic drive in *Drosophila*, the evidence for preferential segregation of chromosomes to a particular pole in the first meiotic division is somewhat stronger (Peacock, 1965; Yanders *et al.*, 1968). In spermatocytes, the *sc*[4]-*sc*[8] X chromosome, deficient for a large heterochromatic segment, frequently fails to pair with the Y chromosome. Cytological observations revealed that in spermatocytes in which the X and Y were unpaired the two chromosomes almost always moved to the same pole in the first meiotic division. Genetic data show that the total frequency of sperm carrying either no sex chromosome, or both the X and Y chromosomes, corresponds well with the frequency of nondisjunction observed cytologically.

The conclusion that the two poles (or the two half-spindles) in the first meiotic division are not functionally equivalent is based on two other remarkable observations. The first is that of all the nondisjunctional sperm transmitted to the progeny the XY class represents only a very small fraction; the great majority of nondisjunctional sperm contain no sex chromosome (Peacock, 1965). Since there is no indication in the genetic data of sperm- or zygote mortality, this observation is best explained by assuming that the unpaired X and Y chromosomes are usually distributed, in the first meiotic division, to a daughter cell that regularly yields nonfunctional sperm. Second, in spermatocytes with an aggregation of bacteria at one of the poles and unpaired sex chromosomes, the X and Y are usually distributed at anaphase to the pole with the bacteria (Yanders *et al.*, 1968).

These observations show that at least in these strains the two poles of the first meiotic spindle differ in a way that is functionally significant in the process of chromosome distribution. What is less certain in this case is that the chromosomes can in any way "recognize" this difference or that the kinetochores of the unpaired sex chromosomes are actively involved. In this connection, it is worth nothing that in the spermatocytes of stone flies unpaired sex chromosomes also regularly pass to the same pole in the first meiotic division, but their appearance suggests that their kinetochores are inactive (Matthey and Aubert, 1947). Previous association in prophase might leave the chromosomes in the same half-spindle after the breakdown of the nuclear envelope, and they would be "eliminated" toward the same pole in the manner of acentric fragments. This kind of passive behavior has been described for the paired sex chromosomes in spermatocytes of the ostracod *Notodromus monacha* (Dietz, 1954). As might be expected if this were the mechanism responsible for the nonrandom distribution of the unpaired *sc*[4]-*sc*[8] X and Y chromosomes in *Drosophila* spermatocytes, the sex chromosomes are usually found at the same pole as the bacteria.

In short, some examples of nonrandom distribution of chromosomes show that there must be a functional difference between the two poles or half spindles; other observations, such as those on the primary spermatocytes of *Gryllotalpa* (described below) and *Sciara,* indicate a functional difference between paired or unpaired homologs during division. Mechanisms of nonrandom distribution based on specific interaction between a particular chromosome and a particular pole require that differences between poles and between chromosomes *both* be present during the division, but this has not yet been conclusively demonstrated for any cell type.

3. Interactions between Chromosomes or Chromosomal Spindle fibers

The observations of Dietz (1969) on tipulids, and of Camenzind and Nicklas (1968) on *Gryllotalpa,* strongly indicate that the mechanism of nonrandom segregation in meiosis in these organisms is based on interactions between chromosomes or chromosomal fibers.

Dietz (1969) has observed that in tipulid spermatocytes with only one sex chromosome this chromosome often remains at the equator throughout division. When two sex chromosomes are present, even though unpaired, they move toward opposite poles at late anaphase. Clearly, then, the opposite segregation of the X and Y chromosomes in normal spermatocytes is determined by an interaction between the two chromosomes or their spindle fibers, not by a specific "attraction" between a chromosome and a particular pole. Dietz proposes that under conditions where microtubule subunits are undergoing solubilization but the microtubules themselves are not free to shorten rapidly—i.e., in the antagonistic kinetochore fibers of sister chromatids that fail to separate in anaphase—the microtubules are subject to an alternate structural change involving a swelling and elongation. This elongation is promoted by the more rapid disassembly and shortening of other microtubules in the vicinity. Chance elongation of one kinetochore fiber of the X chromosome, for example, allows a shortening of the antagonistic kinetochore fiber in the opposite half-spindle; this shortening in turn promotes the elongation of the Y chromosome's kinetochore fiber in the same half-spindle and allows a shortening of the opposite kinetochore fiber of the Y. The two sex chromosomes and their kinetochore fibers thus constitute a system with positive feedback, in which in each half-spindle the kinetochore fiber of one chromosome shortens while that of the other elongates.

Interaction between unpaired chromosomes or their spindle fibers has also been demonstrated in spermatocytes of the mole cricket *Gryllotalpa hexadactyla*. In this organism one of the autosomal bivalents is composed of two unequal chromosomal elements. At metaphase of the first meiotic divison, the bivalent is usually displaced slightly in the direction of the pole toward which the smaller element is oriented. The unpaired X chromosome invariably lies near, and is oriented toward, the other pole. At anaphase the X chromosome and the larger autosomal dyad always move to one pole, the smaller dyad to the opposite pole. Unfortunately, the living cells are not favorable for observations on the prometaphase behavior of the chromosomes in question, but the micromanipulation experiments of Camenzind and Nicklas (1968) have provided evidence that the orientation of the X chromosome is determined by that of the heteromorphic bivalent. If the X chromosome at metaphase is experimentally oriented toward the *same* pole as the smaller half-bivalent (either by inverting the bivalent or by reorienting the X), the X usually reorients toward the opposite pole. Manipulation of the X chromosome, or its removal altogether, has no effect on the orientation of the heteromorphic bivalent. (The observation that reorientations of the X chromosome also occur after *removal* of the heteromorphic bivalent suggests that the bivalent does not merely induce reorientation of the X when the X is oriented toward the same pole as the smaller half-bivalent, but that it contributes in a positive way to the stability of normal X orientation. This conclusion is based, however, on only one cell.)

The physicochemical basis for this control of orientation stability in the absence of physical connections is not known. It probably involves differences between different kinetochores in their ability to form spindle fibers and interactions between the spindle fibers themselves.

Other instances in which the behavior of one or more chromosomes appears to influence decisively the behavior of other physically unconnected chromosomes have been assembled in Table VIII. One great merit of Dietz's hypothesis (Chapter IV, Section D) is that it provides a general mechanism for interactions between spindle fibers of different chromosomes. Nevertheless, without further experimentation it is not clear exactly how this hypothesis would apply to all of the phenomena listed in the table.

TABLE VIII

EXAMPLES OF INTERDEPENDENCE OF MOVEMENTS OF UNPAIRED CHROMOSOMES DURING DIVISION

Phenomenon	Description	Organism	Reference
Distance segregation (= "distance conjugation")	Unpaired X and Y chromosomes move to opposite poles in prometaphase, metaphase, or anaphase	Various liverworts	Lorbeer, 1934
		Various neuropteran insects	Naville and de Beaumont, 1933; Hughes-Schrader, 1969
		Dysdercus koenigii (heteropteran insect)	Ray-Chaudhuri and Manna, 1952
		Various tipulids (dipteran insect)	Dietz, 1969
	X chromosome orients at one pole; six Y chromosomes all orient independently at opposite pole	*Asphaera daniela* (coleopteran insect)	Virkki, 1968
	Central element of trivalent moves to one pole; terminal elements and univalent move to opposite pole	*Austroagalloides* sp. (homopteran insect)	Whitten, 1968
Controlled reorientation and segregation	X chromosome orients toward same pole as larger element of heteromorphic bivalent	*Gryllotalpa* spp.	Camenzind and Nicklas, 1968, and references therein.
	Asynaptic I- and V-shaped univalents orient toward one pole; J-shaped univalent orients toward opposite pole	*Omophoita clerica* (coleopteran insect)	Virkki, 1967
	Nondividing univalents tend to be distributed to the pole toward which fewer kinetochores were oriented at metaphase	Triploid *Anthoxanthum* (angiosperm: Graminales)	Carroll, 1966
		Various tipulids (dipteran insect)	Dietz, 1969

Interdependence of anaphase movements	Slowing or stopping of dyad by UV microbeam irradiation in one half-spindle is always associated with slowing or stopping of homologous dyad in the unirradiated half-spindle	*Nephrotoma suturalis* (dipteran insect)	Forer, 1966
Timing of onset of anaphase	Anaphase begins for all bivalents only after the unpaired X chromosome has completed poleward migration	*Melanoplus differentialis* (orthopteran insect)	Nicklas, 1961
	Anaphase begins only after a certain number of univalents have accumulated at the equator of the spindle	*Triticum timopheevi* × *T. durum* hybrid (angiosperm: Graminales)	Wagenaar, 1961a

F. SUMMARY

A "two-component" model of chromosome orientation and movement has been presented in which poleward-oriented, force-generating elements act on noncontractile, labile, polarized kinetochore microtubules, transmitting a poleward pulling force to the kinetochore. The net pulling force on a kinetochore is the resultant of several independent factors: the number and length of kinetochore microtubules, the relative orientation of microtubules and force-generating elements, and the activity of the kinetochore in directing the assembly of microtubules. This model has been contrasted with Dietz's "one-component" model, in which chromosome movement is caused by changes within the kinetochore microtubules themselves. Neither model accounts fully for all observations on movements occurring during cell division, and there are no observations at present that can critically distinguish between the two models.

Examination of several cases of nonrandom chromosome segregation suggests that in some cells there are differences between chromosomes with respect to properties affecting their orientation and movement, and that in some cells there are functional differences between the two poles of the spindle (or between the two half-spindles). In no single case is there convincing evidence that functional differences between chromosomes and between poles coexist in the same cell. Cases of nonrandom segregation can be explained in terms of mitotic forces already known or in terms of interactions between different chromosomes or their kinetochore fibers; they require the assumption neither of persistent connections between chromosomes and poles, nor of specific interactions between a particular chromosome and a particular pole.

ACKNOWLEDGMENTS

This review was supported by grants from the U.S. Public Health Service (NIH HD-03519 and -00187).

REFERENCES

Abuelo, J. G., and Moore, D. E. (1969). *J. Cell Biol.* **41,** 73.

Aikawa, M., and Beaudoin, R. L. (1968). *J. Cell Biol.* **39,** 749.

Aikawa, M., Huff, C. G., and Sprinz, H. (1967). *J. Cell Biol.* **34,** 229.

Aldrich, H. C. (1967). *Mycologia* **59,** 127.

Aldrich, H. C. (1969). *Am. J. Botany* **56,** 290.

Allen, M. M., and Stanier, R. Y. (1968). *J. Gen. Microbiol.* **51,** 199.

Allen, R. D. (1969). *J. Cell Biol.* **40,** 716.

Allen, R. D., and Bowen, C. C. (1966). *Caryologia* **19,** 299.

Allenspach, A. L., and Roth, L. E. (1967). *J. Cell Biol.* **33,** 179.

Bajer, A. (1958a). *Chromosoma* **9,** 319.

Bajer, A. (1958b). *Exptl. Cell Res.* **14,** 245.

Bajer, A. (1958c). *Exptl. Cell Res.* **15,** 370.

Bajer, A. (1961). *Chromosoma* **12,** 64.

Bajer, A. (1963). *Chromosoma* **14,** 18.

Bajer, A. (1964). *Chromosoma* **15,** 630.

Bajer, A. (1966a). *In* "Probleme der biologischen Reduplikation" (P. Sitte, ed.), pp. 90–119. Springer, Berlin.

Bajer, A. (1966b). *In* "Dynamics of Fluids and Plasmas" (S. I. Pai, ed.), pp. 59–80. Academic Press, New York.

Bajer, A. (1967). *J. Cell Biol.* **33,** 713.

Bajer, A. (1968a). *In* "Aspects of Cell Motility" (22nd Symp. Soc. Exptl. Biol.), pp. 286–310. Columbia Univ. Press, New York.

Bajer, A. (1968b). *Chromosoma* **25,** 249.

Bajer, A. (1968c). *Chromosoma* **24,** 383.

Bajer, A. (1969). *Biophys. J.* **9,** A-151.

Bajer, A., and Molè-Bajer, J. (1956). *Chromosoma* **7,** 558.

Bajer, A., and Molè-Bajer, J. (1961). *Exptl. Cell Res.* **25,** 251.

Bajer, A., and Molè-Bajer, J. (1962). Mitosis in Endosperm. III B. 16-mm film. K. Gluchowski, 66 Hartswood Road, London W.12, England.

Bajer, A., and Molè-Bajer, J. (1963). *In* "Cinemicrography in Cell Biology" (G. C. Rose, ed.), pp. 357–409. Academic Press, New York.

Bajer, A., and Molè-Bajer, J. (1969). *Chromosoma* **27,** 448.

Baltzer, F. (1908). *Verhandl. Physik. Med. Ges., Würzburg* [N.F.] **39,** 291.

Bannister, L. H., and Tatchell, E. C. (1968). *J. Cell Sci.* **3,** 295.

Barber, H. N. (1940). *Proc. Roy. Soc.* **B128,** 170.

Barber, H. N. (1941). *J. Genet.* **42,** 223.

Barnicot, N. A. (1966). *J. Cell Sci.* **1,** 217.

Barnicot, N. A., and Huxley, H. E. (1965). *Quart. J. Microscop. Sci.* **106,** 197.

Barski, G., Sorieul, S., and Cornefert, F. (1961). *J. Natl. Cancer Inst.* **26,** 1269.

Barthelmess, A. (1957). *Protoplasma* **48,** 546.

Barthelmess, A., and Einlechner, J. (1959). *Protoplasma* **51,** 325.

Barthelmess, A., and Elkabarity, A. (1961). *Protoplasma* **54,** 455.

Bauer, H. (1967). *Chromosoma* **22,** 101.

Bauer, H., Dietz, R., and Röbbelen, C. (1961). *Chromosoma* **12,** 116.

Bawden, F. C., Pirie, N. W., Bernal, J. D., and Fankuchen, I. (1936). *Nature* **138,** 1051.

Beachey, E. H., and Cole, R. M. (1966). *J. Bacteriol.* **92,** 1245.

Beermann, S. (1959). *Chromosoma* **10,** 504.

Beermann, S. (1966). *Genetics* **54,** 567.

Behnke, O., and Forer, A. (1966). *Compt. Rend. Trav. Lab. Carlsberg* **35,** 437.

Behnke, O., and Forer, A. (1967). *J. Cell Sci.,* **2,** 169.

Beisson, J., and Sonneborn, T. M. (1965). *Proc. Natl. Acad. Sci. U.S.* **53,** 275.

Bělař, K. (1929). *Arch. Entwicklungsmech. Organ.* **118,** 359.

Bělař, K., and Huth, W. (1933). *Z. Zellforsch. Mikroskop. Anat.* **17,** 51.

Bendich, A., and Rosenkranz, H. S. (1963). *Progr. Nucleic Acid Res.* **1,** 219.

Benoit, J., and Kehl, R. (1939). *Compt. Rend. Soc. Biol.* **131,** 329.

Bernal, J. D. (1940). *Am. Assoc. Advan. Sci. Publ.* **14,** 199.

Bernal, J. D., and Fankuchen, I. (1941). *J. Gen. Physiol.* **25,** 111.

Bernhard, W., and de Harven, E. (1960). *In 4th Intern. Congr. Electron Microscop. Berlin, 1958* Part II, pp. 217–227. Springer, Berlin.

Bessis, M., Breton-Gorius, J., and Thiéry, J. P. (1958). *Rev. Hematol.* **13,** 363.

Bevan, P., and Douglas, H. C. (1969). *J. Bacteriol.* **98,** 532.

Bianchi, N. O., and de Bianchi, M. S. A. (1965). *Chromosoma* **17,** 273.

Bibring, T. (1962). Ph.D. Thesis, Univ. of California, Berkeley, California.

Bibring, T., and Baxandall, J. (1968). *Science* **161,** 377.

Bikle, D., Tilney, L. G., and Porter, K. R. (1966). *Protoplasma* **61,** 322.

Birnstiel, M. L., Wallace, H., Sirlin, J. L., and Fischberg, M. (1965). *Natl. Cancer Inst. Monograph* **23,** 431.

Bloom, A. D., Archer, P. G., and Awa, A. A. (1967). *Nature* **216,** 487.

Bloom, W., Zirkle, R. E., and Uretz, R. B. (1955). *Ann. N.Y. Acad. Sci.* **59,** 503.

Borenfreund, E., Krim, M., and Bendich, A. (1964). *J. Natl. Cancer Inst.* **32,** 667.

Borisy, G. G., and Taylor, E. W. (1967a). *J. Cell Biol.* **34,** 525.

Borisy, G. G., and Taylor, E. W. (1967b). *J. Cell Biol.* **34,** 535.

Boss, J. (1954). *Exptl. Cell Res.* **7,** 215.

Boveri, M. (1903). *Jena. Z. Naturw.* **37,** 401.

Boveri, T. (1900). "Zellen-Studien," Part IV: Ueber die Natur der Centrosomen. Fischer, Jena.

Boveri, T. (1907). "Zellen-Studien," Part VI: Die Entwicklung dispermer Seeigel-Eier. Ein Beitrag zur Befruchtungslehre und zur Theorie des Kerns. Fischer, Jena.

Bowers, B., and Korn, E. D. (1968). *J. Cell Biol.* **39,** 95.

Bowman, J. E., Carson, P. E., Frischer, H., and de Garay, A. L. (1966). *Nature* **210,** 811.

Boyer, S. H., Porter, I. H., and Weilbacher, R. G. (1962). *Proc. Natl. Acad. Sci. U.S.* **48,** 1868.

Brachet, J. (1955). *Biochim. Biophys. Acta* **18,** 247.

Brachet, J. (1957). "Biochemical Cytology." Academic Press, New York.

Brachet, J., and Chantrenne, H. (1956). *Cold Spring Harbor Symp. Quant. Biol.* **21,** 329.

Brandham, P. E. (1969). *Chromosoma* **26,** 270.

Brinkley, B. R., and Stubblefield, E. (1966). *Chromosoma* **19,** 28.

Brinkley, B. R., Stubblefield, E., and Hsu, T. C. (1967). *J. Ultrastruct. Res.* **19,** 1.
Brown, S. W. (1966). *Science* **151,** 417.
Brown, S. W., and Zohary, D. (1955). *Genetics* **40,** 850.
Buck, R. C. (1967). *J. Ultrastruct. Res.* **18,** 489.
Burton, P. R., and Dusanic, D. G. (1968). *J. Cell Biol.* **39,** 318.
Byers, B., and Abramson, D. H. (1968). *Protoplasma* **66,** 413.
Callan, H. G., and Jacobs, P. (1957). *J. Genet.* **55,** 200.
Callan, H. G., and Taylor, J. H. (1968). *J. Cell Sci.* **3,** 615.
Camenzind, R., and Nicklas, R. B. (1968). *Chromosoma* **24,** 324.
Campbell, R. D., and Inoué, S. (1965). *Biol. Bull.* **129,** 401.
Carasso, N., and Favard, P. (1961). *In* "Traité de Microscopie Électronique" (C. Magnan, ed.), Vol. II, pp. 901–1117. Hermann, Paris.
Carasso, N., and Favard, P. (1965). *J. Microscop.* Paris **4,** 395.
Carlson, J. G. (1938). *Proc. Natl. Acad. Sci. U.S.* **24,** 500.
Carlson, J. G. (1956). *Science* **124,** 203.
Carroll, C. P. (1966). *Chromosoma* **18,** 19.
Caspar, D. L. D., and Klug, A. (1962). *Cold Spring Harbor Symp. Quant. Biol.* **27,** 1.
Chai, N. -C., and Lark, K. G. (1967). *J. Bacteriol.* **94,** 415.
Chambers, R., Jr. (1917). *J. Exptl. Zool.* **23,** 483.
Charret, R., and André, J. (1968). *J. Cell Biol.* **39,** 369.
Clark, F. J. (1940). *Am. J. Botany* **27,** 547.
Cleveland, L. R. (1935). *Biol. Bull.* **69,** 46.
Cleveland, L. R. (1953). *Trans. Am. Phil. Soc.* **43,** 809.
Cleveland, L. R. (1958a). *J. Protozool.* **5,** 47.
Cleveland, L. R. (1958b). *J. Protozool.* **5,** 63.
Cleveland, L. R. (1960). *J. Protozool.* **7,** 326.
Cloney, R. A. (1966). *J. Ultrastruct. Res.* **14,** 300.
Cole, R. M. (1965). *Bacteriol. Rev.* **29,** 326.
Comings, D. E. (1966). *Science* **154,** 1463.
Comings, D. E. (1968). *Am. J. Human Genet.* **20,** 440.
Conger, A. D. (1940). *J. Heredity* **31,** 339.
Conklin, E. G. (1904). *Biol. Bull.* **7,** 221.
Cooper, K. W. (1951). *J. Morphol.* **89,** 37.
Cooper, K. W. (1964). *Proc. Natl. Acad. Sci. U.S.* **52,** 1248.
Cooper, S., and Helmstetter, C. E. (1968). *J. Mol. Biol.* **31,** 519.
Costello, D. P. (1961). *Biol. Bull.* **120,** 285.
Crick, F. H. C., and Hughes, A. F. W. (1950). *Exptl. Cell Res.* **1,** 37.
Cronshaw, J., and Esau, K. (1967). *J. Cell Biol.* **34,** 801.
Cronshaw, J., and Esau, K. (1968). *Protoplasma* **65,** 1.
Crouse, H. V. (1960). *Genetics* **45,** 1429.
Cuevas-Sosa, A. (1968). *Nature* **218,** 1059.
Cuzin, F., and Jacob, F. (1965). *Compt. Rend. Acad. Sci.* **260,** 5411.
Daker, M. G. (1967). *Chromosoma* **21,** 250.
Daniels, E. W., and Roth, L. E. (1964). *J. Cell Biol.* **20,** 75.
Darlington, C. D. (1939). *J. Genet.* **37,** 341.
Darlington, C. D. (1940). *Biol. Rev. Cambridge Phil. Soc.* **15,** 307.
Darlington, C. D. (1965). "Cytology." Churchill, London.

Darlington, C. D., and Dark, S. O. S. (1932). *Cytologia* **3,** 169.
Darlington, C. D., and Thomas, P. T. (1937). *Ann. Botany (London)* **1,** 747.
Darlington, C. D., and Vosa, C. G. (1963). *Chromosoma* **13,** 609.
Dass, C. M. S., and Sapra, G. R. (1966). *Exptl. Cell Res.* **43,** 622.
David, P. R. (1936). *Arch. Entwicklungsmech. Organ.* **135,** 521.
Davidson, E. H. (1964). *Advan. Genet.* **12,** 143.
Davidson, R. G., and Cortner, J. A. (1967). *Science* **157,** 1569.
Davis, B. M. (1901). *Ann. Botany (London)* **15,** 147.
de Harven, E. (1968). *In* "The Nucleus" (A. J. Dalton and F. Hagenau, eds.), pp. 197–227. Academic Press, New York.
DeLorenzo, R. J., and Ruddle, F. H. (1969). *Biochem. Genet.* **3,** 151.
Demerec, M. (1934). *Proc. Natl. Acad. Sci. U.S.* **20,** 354.
Demerec, M. (1936). *Proc. Natl. Acad. Sci. U.S.* **22,** 350.
Denell, R. E., and Judd, B. H. (1969). *Molec. Gen. Genetics* **105,** 262.
de-Thé, G. (1964). *J. Cell Biol.* **23,** 265.
de Torok, D., and White, P. R. (1960). *Science* **131,** 730.
Detter, J. C., Ways, P. O., Giblett, E. R., Baughan, M. A., Hopkinson, D. A., Povey, S., and Harris, H. (1968). *Ann. Human Genet.* **31,** 329.
Devisé, R. (1922). *Cellule* **32,** 249.
Dietrich, J. (1966). *Compt. Rend. Acad. Sci. Ser. D* **262,** 1699.
Dietrich, J. (1968). *Compt. Rend. Acad. Sci. Ser D* **266,** 579.
Dietz, R. (1954). *Chromosoma* **6,** 397.
Dietz, R. (1956). *Chromosoma* **8,** 183.
Dietz, R. (1958). *Chromosoma* **9,** 359.
Dietz, R. (1959). *Z. Naturforsch.* **14b,** 749.
Dietz, R. (1963). *Zool. Anz. Suppl.* **26,** 131.
Dietz, R. (1966). *In* "Chromosomes Today" (C. D. Darlington and K. R. Lewis, eds.), Vol. 1, pp. 161–166. Oliver and Boyd, Edinburgh.
Dietz, R. (1969). *Naturwissenschaften* **56,** 237.
Dingle, A. D., and Fulton, C. (1966). *J. Cell Biol.* **31,** 43.
Dippell, R. V. (1968). *Proc. Natl. Acad. Sci. U.S.* **61,** 461.
Dirksen, E. R. (1961). *J. Biophys. Biochem. Cytol.* **11,** 244.
Dowrick, G. J. (1957). *Heredity* **11,** 37.
Drum, R. W., and Pankratz, H. S. (1963). *Science* **142,** 61.
Duncan, R. E., and Persidsky, M. D. (1958). *Am. J. Botany* **45,** 719.
DuPraw, E. J. (1966). *Nature* **209,** 577.
DuPraw, E. J. (1968). "Cell and Molecular Biology." Academic Press, New York.
Eberle, H., and Lark, K. G. (1966). *J. Mol. Biol.* **22,** 183.
Ellar, D. J., Lundgren, D. G., and Slepecky, R. A. (1967). *J. Bacteriol.* **94,** 1189.
Ephrussi, B. (1934). *Proc. Natl. Acad. Sci. U.S.* **20,** 420.
Ephrussi, B. (1935). *J. Exptl. Zool.* **70,** 197.
Erickson, J., and Acton, A. B. (1969). *Can. J. Genet. Cytol.* **11,** 153.
Fabbri, F. (1960). *Caryologia* **13,** 297.
Fankhauser, G. (1934a). *J. Exptl. Zool.* **67,** 349.
Fankhauser, G. (1934b). *J. Exptl. Zool.* **68,** 1.
Fankhauser, G., and Humphrey, R. R. (1950). *J. Exptl. Zool.* **115,** 207.
Fankhauser, G., and Humphrey, R. R. (1959). *J. Exptl. Zool.* **142,** 379.

Fell, H. B., and Hughes, A. F. (1949). *Quart. J. Microscop. Sci.* **90,** 355.
Ferguson, M. C. (1901a). *Ann. Botany (London)* **15,** 193.
Ferguson, M. C. (1901b). *Ann. Botany (London)* **15,** 435.
Fetner, R. H., and Porter, E. D. (1965). *Exptl. Cell Res.* **37,** 429.
Fialkow, P. J., Lisker, R., Detter, J., Giblett, E. R., and Zavala, C. (1969). *Science* **163,** 194.
Fildes, R. A., and Parr, C. W. (1963). *Nature* **200,** 890.
Ford, C. E. (1964). *Symp. Intern. Soc. Cell Biol.* **3,** 27–45.
Forer, A. (1965). *J. Cell Biol.* **25,** 95.
Forer, A. (1966). *Chromosoma* **19,** 44.
Fredga, K. (1967). *Hereditas* **57,** 421.
Frew, P. E., and Bowen, R. H. (1929). *Quart. J. Microscop. Sci.* **73,** 197.
Friedländer, M., and Wahrman, J. (1966). *J. Cell Sci.* **1,** 129.
Fuhs, G. W. (1969). *Protoplasmatologia* **V4.**
Gabriel, M. L. (1960). *Am. Naturalist* **94,** 257.
Gall, J. G. (1961). *J. Biophys. Biochem. Cytol.* **10,** 163.
Ganesan, A. T., and Lederberg, J. (1965). *Biochem. Biophys. Res. Commun.* **18,** 824.
Gavin, R. H., and Frankel, J. (1966). *J. Exptl. Zool.* **161,** 63.
George, P., Journey, L. J., and Goldstein, M. M. J. (1965). *J. Natl. Cancer Inst.* **35,** 355.
German, J. (1966). *Chromosoma* **20,** 195.
Gibbins, J. R., Tilney, L. G., and Porter, K. R. (1969a). *J. Cell Biol.* **41,** 201.
Gibbins, J. R., Tilney, L. G., and Porter, K. R. (1969b). *J. Cell Biol.* **41,** 227.
Giménez-Martín, G., González-Fernández, A., and López-Sáez, J. F. (1964). *Phyton* **21,** 77.
Gläss, E. (1961). *Chromosoma* **12,** 422.
Gläss, E. (1962). *Experientia* **18,** 192.
Goldman, R. D., and Rebhun, L. I. (1969). *J. Cell Sci.* **4,** 179.
Goldstein, L. (1967). *Science* **156,** 1133.
Gonatas, N. K., and Robbins, E. (1965). *Protoplasma* **59,** 377.
Gonzalez, M. A. (1967). *Genet. Iberica* **19,** 1.
González-Fernández, A., López-Sáez, J. F., and Giménez-Martín, G. (1964). *Phyton* **21,** 157.
Gould, M. C. (1968). *J. Cell Biol.* **38,** 220.
Grell, E. H. (1967). *Science* **158,** 1319.
Grell, K. G. (1953). *Arch. Protistenk.* **99,** 1.
Grell, K. G., and Ruthmann, A. (1964). *Chromosoma* **15,** 185.
Grell, K. G., and Schwalbach, G. (1965). *Chromosoma* **17,** 230.
Grimstone, A. V. (1967). *Symp. Intern. Soc. Cell Biol.* **6,** 219–232.
Grimstone, A. V., and Cleveland, L. R. (1965). *J. Cell Biol.* **24,** 387.
Grimstone, A. V., and Gibbons, I. R. (1966). *Phil. Trans. Roy. Soc. London Ser. B* **250,** 215.
Gülcan, R., and Sybenga, J. (1967). *Genetica* **38,** 163.
Guttes, S., Guttes, E., and Ellis, R. A. (1968). *J. Ultrastruct. Res.* **22,** 508.
Hadorn, E. (1961). "Developmental Genetics and Lethal Factors." Wiley, New York.
Hanawalt, P. C., and Ray, D. S. (1964). *Proc. Natl. Acad. Sci. U.S.* **52,** 125.

Hanson, E. D. (1967). *In* "Chemical Zoology" (M. Florkin and B. Scheer, eds.), Vol. I (Protozoa), pp. 395–539. Academic Press, New York.

Hanson, J., and Lowy, J. (1963). *J. Mol. Biol.* **6,** 46.

Harris, H., Watkins, J. F., Ford, C. E., and Schoefl, G. I. (1966). *J. Cell Sci.* **1,** 1.

Harris, J. W. (1950). *Proc. Soc. Exptl. Biol. Med.* **75,** 197.

Harris, M. (1964). "Cell Culture and Somatic Variation." Holt, New York.

Harris, P. (1961). *J. Biophys. Biochem. Cytol.* **11,** 419.

Harris, P. (1962a). *In 5th Intern. Congr. Electron Microscopy.* p. NN-1. Academic Press, New York.

Harris, P. (1962b). *J. Cell Biol.* **14,** 475.

Harris, P. (1965). *J. Cell Biol.* **25,** 73.

Harris, P., and Bajer, A. (1965). *Chromosoma* **16,** 624.

Harris, P., and Mazia, D. (1962). *Symp. Intern. Soc. Cell Biol.* **1,** 279–305.

Hartl, D. L. (1969). *Proc. Natl. Acad. Sci. U.S.* **63,** 782.

Hartl, D. L., Hiraizumi, Y., and Crow, J. F. (1967). *Proc. Natl. Acad. Sci. U.S.* **58,** 2240.

Hartmann, M. (1933). "Allgemeine Biologie," 2nd Ed. Fischer, Jena.

Hartwell, L. H., and McLaughlin, C. S. (1968). *Proc. Natl. Acad. Sci. U.S.* **59,** 422.

Hartwell, L. H., and McLaughlin, C. S. (1969). *Proc. Natl. Acad. Sci. U.S.* **62,** 468.

Hayflick, L., and Moorhead, P. S. (1961). *Exptl. Cell Res.* **25,** 585.

Hearne, E. M., and Huskins, C. L. (1935). *Cytologia* **6,** 123.

Heath, I. B., and Greenwood, A. D. (1968). *J. Gen. Microbiol.* **53,** 287.

Heddle, J. A., Wolff, S., Whissel, D., and Cleaver, J. E. (1967). *Science* **158,** 929.

Henderson, N. S. (1965). *J. Exptl. Zool.* **158,** 263.

Henderson, N. S. (1966). *Arch. Biochem. Biophys.* **117,** 28.

Henderson, S. A. (1962). *Chromosoma* **13,** 437.

Henderson, S. A. (1965). *Chromosoma* **16,** 192.

Henderson, S. A. (1967). *Chromosoma* **23,** 38.

Heneen, W. K., Nichols, W. W., Levan, A., and Norrby, E. (1967). *Hereditas* **57,** 369.

Henshaw, P. S. (1940). *Am. J. Roentgenol. Radium Therapy Nucl. Med.* **43,** 923.

Hepler, P. K., and Jackson, W. T. (1968). *J. Cell Biol.* **38,** 437.

Hepler, P. K., Huff, C. G., and Sprinz, H. (1966). *J. Cell Biol.* **30,** 333.

Herreros, B., and Gianelli, F. (1967). *Nature* **216,** 286.

Hertwig, R. (1899). *Abhandl. Math.-Phys. Kl. Königliche Bayerischen Akad. Wiss.* **19,** 631.

Heslop-Harrison, J. (1968). *J. Cell Sci.* **3,** 457.

Highton, P. J. (1969). *J. Ultrastruct. Res.* **26,** 130.

Hoffman, E. J. (1965). *J. Cell. Biol* **25,** 217.

Hollande, A., and Valentin, J. (1967). *Compt. Rend. Acad. Sci. Ser. D* **264,** 1868.

Hollande, A., and Valentin, J. (1968a). *Compt. Rend. Acad. Sci. Ser D* **266,** 367.

Hollande, A., and Valentin, J. (1968b). *Compt. Rend. Acad. Sci. Ser. D* **266,** 1283.

Hollande, A., and Valentin, J. (1968c). *Compt. Rend. Acad. Sci. Ser. D* **267,** 1383.

Hollande, A., Cachon, J., and Cachon, M. (1969). *Compt. Rend. Acad. Sci. Ser. D* **269,** 179.

Holm, G., and Bajer, A. (1966). *Hereditas* **54,** 356.

Holmes, K. V., and Choppin, P. W. (1968). *J. Cell Biol.* **39,** 526.

Hopkinson, D. A., and Harris, H. (1968). *Ann. Human Genet.* **31,** 359.
Hsu, T. C. (1961). *Intern. Rev. Cytol.* **12,** 69.
Huettner, A. F. (1933). *Z. Zellforsch. Mikroskop. Anat.* **19,** 119.
Hughes, A. F., and Swann, M. M. (1948). *J. Exptl. Biol.* **25,** 45.
Hughes-Schrader, S. (1924). *J. Morphol.* **39,** 157.
Hughes-Schrader, S. (1935). *Biol. Bull.* **69,** 462.
Hughes-Schrader, S. (1942). *J. Morphol.* **70,** 261.
Hughes-Schrader, S. (1943). *Biol. Bull.* **85,** 265.
Hughes-Schrader, S. (1948a). *Chromosoma* **3,** 257.
Hughes-Schrader, S. (1948b). *Advan. Genet.* **2,** 127.
Hughes-Schrader, S. (1950). *Chromosoma* **4,** 1.
Hughes-Schrader, S. (1969). *Chromosoma* **27,** 109.
Hungerford, D. A., Makino, S., Sasaki, M., Awa, A. A., and Balaban, G. B. (1969). *Cytogenetics* **8,** 74.
Huxley, H. E. (1963). *J. Mol. Biol.* **7,** 281.
Ichida, A. A., and Fuller, M. S. (1968). *Mycologia* **60,** 141.
Inamdar, N. B. (1949). *Biol. Bull.* **97,** 300.
Inoué, S. (1953). *Chromosoma* **5,** 487.
Inoué, S. (1964). *In* "Primitive Motile Systems in Cell Biology" (R. D. Allen and N. Kamiya, eds.), pp. 549–598. Academic Press, New York.
Inoué, S., and Bajer, A. (1961). *Chromosoma* **12,** 48.
Inoué, S., and Sato, H. (1967). *J. Gen. Physiol.* **50,** 259.
Ito, M., Hotta, Y., and Stern, H. (1967). *Develop. Biol.* **16,** 54.
Ito, S. (1960). *J. Biophys. Biochem. Cytol.* **7,** 433.
Iwata, J. (1940). *Japan. J. Botany* **10,** 375.
Izutsu, K. (1960). *Cytologia* **25,** 293.
Izutsu, K. (1961). *Mie Med. J.* **11,** 213.
Jackson, W. D., and Cheung, D. S. M. (1967). *Chromosoma* **23,** 24.
Jacob, F., Brenner, S., and Cuzin, F. (1963). *Cold Spring Harbor Symp. Quant. Biol* **28,** 329.
Jacob, F., Ryter, A., and Cuzin, F. (1966). *Proc. Roy. Soc.* **B164,** 267.
Jacquez, J. A., and Biesele, J. J. (1954). *Exptl. Cell Res.* **6,** 17.
Jain, H. K. (1957). *Heredity* **11,** 23.
Jenkins, R. A. (1967). *J. Cell Biol.* **34,** 463.
Jensen, C., and Bajer, A. (1969). *J. Ultrastruct. Res.* **26,** 367.
John, B. (1957a). *Chromosoma* **8,** 730.
John, B. (1957b). *Heredity* **11,** 209.
John, B., and Hewitt, G. M. (1965). *Chromosoma* **16,** 548.
John, B., and Hewitt, G. M. (1966). *Chromosoma* **20,** 155.
John, B., and Hewitt, G. M. (1968). *Chromosoma* **25,** 40.
John, B., and Lewis, K. R. (1960). *Heredity* **15,** 431.
John, B., and Lewis, K. R. (1965). *Protoplasmatologia* **VI F1.**
Johnson, H. H. (1931). *Z. Wiss. Zool.* **140,** 115.
Johnson, U. G., and Porter, K. R. (1968). *J. Cell Biol.* **38,** 403.
Jokelainen, P. T. (1965). *J. Cell Biol.* **27,** 48A.
Jokelainen, P. T. (1967). *J. Ultrastruct. Res.* **19,** 19.
Jokelainen, P. T. (1968). *J. Cell Biol.* **39,** 68A.

Jones, K., and Colden, C. (1968). *Chromosoma* **24,** 135.
Jordan, S. W. (1964). *Exptl. Mol. Pathol.* **3,** 76.
Juel, H. O. (1897). *Jahrb. Wiss. Botan.* **30,** 205.
Käfer, E. (1961). *Genetics* **46,** 1581.
Kane, R. E., and Forer, A. (1965). *J. Cell Biol.* **25,** 31.
Kao, F. -T., and Puck, T. T. (1967). *Genetics* **55,** 513.
Kato, R. (1968). *Hereditas* **59,** 63.
Kaylor, C. T. (1941). *Biol. Bull.* **81,** 402.
Kemp, C. L. (1964). *Chromosoma* **15,** 652.
Kessel, R. G. (1966). *J. Ultrastruct. Res.* **16,** 293.
Keyl, H. -G., and Hägele, K. (1966). *Chromosoma* **19,** 223.
Kiefer, B., Sakai, H., Solari, A. J., and Mazia, D. (1966). *J. Mol. Biol.* **20,** 75.
Kimber, G., and Riley, R. (1963). *Botan. Rev.* **29,** 480.
Klingstedt, H. (1938). *Nature* **141,** 606.
Klingstedt, H. (1939). *J. Genet.* **37,** 391.
Kodani, M., and Stern, C. (1946). *Science* **104,** 620.
Kojima, M. K. (1960). *Embryologia* **5,** 71.
Koller, P. C. (1938). *J. Genet.* **36,** 275.
Koopmans, A. (1958). *Naturwissenschaften* **45,** 66.
Krishan, A. (1968). *J. Ultrastruct. Res.* **23,** 134.
Krishan, A., and Buck, R. C. (1965). *J. Cell Biol.* **24,** 433.
Kubai, D. F., and Ris, H. (1969). *J. Cell Biol.* **40,** 508.
Lacadena, J. R., and Ramos, A. (1968). *Genet. Iberica* **20,** 55.
Lark, K. G. (1966). *Bacteriol. Rev.* **30,** 3.
Lark, K. G. (1967). *Proc. Natl. Acad. Sci. U.S.* **58,** 352.
Lark, K. G. (1969). *Genetics* **62,** 289.
Lark, K. G., and Bird, R. E. (1965). *Proc. Natl. Acad. Sci. U.S.* **54,** 1444.
Lark, K. G., Consigli, R. A., and Minocha, H. C. (1966). *Science* **154,** 1202.
Ledbetter, M. C., and Porter, K. R. (1963). *J. Cell Biol.* **19,** 239.
Lefevre, G. (1907). *J. Exptl. Zool.* **4,** 91.
Lentz, T. L. (1967). *J. Ultrastruct. Res.* **17,** 114.
Lerbs, V., and Thielke, C. (1969). *Arch. Mikrobiol.* **68,** 95.
Lessie, P. E., and Lovett, J. S. (1968). *Am. J. Botany* **55,** 220.
Lettré, H., and Lettré, R. (1958). *Rev. Hematol.* **13,** 337.
Lettré, H., and Lettré, R. (1959). *Nucleus* **2,** 23.
Lettré, H., and Lettré, R. (1960). *In Colloq. Intern. Centre Rech. Sci.* **88,** 25–50.
Levan, A., Fredga, K., and Sandberg, A. A. (1964). *Hereditas* **52,** 201.
Levis, A. G., and Marin, G. (1963). *Exptl. Cell Res.* **31,** 448.
Lewis, K. R., and John, B. (1963). *Chromosoma* **14,** 618.
Li, J.-C. (1927). *Genetics* **12,** 1.
Lima-de-Faria, A. (1955a). *Chromosoma* **7,** 51.
Lima-de-Faria, A. (1955b). *Hereditas* **41,** 238.
Lima-de-Faria, A. (1956). *Hereditas* **42,** 85.
Lima-de-Faria, A. (1958a). *Cytologia* **7,** 123.
Lima-de-Faria, A. (1958b). *J. Heredity* **49,** 299.
Lima-de-Faria, A., and Bose, S. (1962). *Chromosoma* **13,** 315.
Lindsley, D. L., and Novitski, E. (1958). *Genetics* **43,** 790.

Longevialle, M. (1968). *Compt. Rend. Acad. Sci. Ser. D* **267,** 708.
Longo, F. J., and Anderson, E. (1968). *J. Cell Biol.* **39,** 339.
Longo, G. P., and Scandalios, J. G. (1969). *Proc. Natl. Acad. Sci. U.S.* **62,** 104.
Lorbeer, G. (1934). *Jahrb. Wiss. Botan.* **80,** 567.
Lorch, I. J. (1952). *Quart. J. Microscop. Sci.* **93,** 475.
Lu, B. C. (1967). *J. Cell Sci.* **2,** 529.
Lubs, H. A., and Blitman, S. L. (1967). *Experientia* **23,** 1067.
Luykx, P. (1965a). *Exptl. Cell Res.* **39,** 643.
Luykx, P. (1965b). *Exptl. Cell Res.* **39,** 658.
McClintock, B. (1938). *Genetics* **23,** 315.
McConkey, E. H., and Hopkins, J. W. (1964). *Proc. Natl. Acad. Sci. U.S.* **51,** 1197.
MacDonald, K. B., and Bruce, W. R. (1968). *Exptl. Cell Res.* **50,** 471.
MacGregor, H. C., and Callan, H. G. (1962). *Quart. J. Microscop. Sci.* **103,** 173.
McIntosh, J. R., and Porter, K. R. (1967). *J. Cell Biol.* **35,** 153.
McIntosh, J. R., Hepler, P. K., and van Wie, D. G. (1969). *Nature* **224,** 659.
McLaughlin, C. S., and Hartwell, L. H. (1969). *Genetics* **61,** 557.
McManus, Sr. M. A., and Roth, L. E. (1965). *J. Cell Biol.* **25,** 305.
McManus, Sr. M. A., and Roth, L. E. (1968). *Mycologia* **60,** 426.
Makino, S., and Nakanishi, Y. H. (1955). *Chromosoma* **7,** 439.
Malawista, S. E., Asterita, H., and Marsland, D. (1966). *J. Cell Physiol.* **68,** 13.
Malawista, S. E., Sato, H., and Bensch, K. G. (1968). *Science* **160,** 770.
Manton, I. (1964a). *Proc. Roy. Microscop. Soc.* **83,** 47.
Manton, I. (1964b). *J. Roy. Microscop. Soc.* **83,** 317.
Manton, I., Kowallik, K., and von Stosch, A. (1969). *J. Cell Sci.* **5,** 271.
Mark, J. (1967). *Hereditas* **57,** 23.
Marr, A. G., Harvey, R. J., and Trentini, W. C. (1966). *J. Bacteriol.* **91,** 2388.
Marsland, D., and Meisner, D. (1967). *J. Cell Physiol.* **70,** 209.
Martini, G., and Bozzini, A. (1966). *Chromosoma* **20,** 251.
Mason, D. J., and Powelson, D. M. (1956). *J. Bacteriol.* **71,** 474.
Matthey, R., and Aubert, J. (1947). *Bull. Biol. France Belg.* **81,** 202.
Matuszewski, B. (1962). *Chromosoma* **12,** 741.
Matuszewski, B. (1966). *Chromosoma* **19,** 194.
Mazia, D. (1961). *In* "The Cell" (J. Brachet and A. E. Mirsky, eds.), Vol. III, pp. 77–412. Academic Press, New York.
Mazia, D. (1967). *Symp. Intern. Soc. Cell Biol.* **6,** 39–54.
Mazia, D., Harris, P., and Bibring, T. (1960). *J. Biophys. Biochem. Cytol.* **7,** 1.
Mead, A. D. (1898). *J. Morphol.* **14,** 181.
Metz, C. W. (1933). *Biol. Bull.* **64,** 333.
Metz, C. W. (1938). *Am. Naturalist* **72,** 485.
Metz, C. W., Moses, M. S., and Hoppe, E. N. (1926). *Z. Induktive Acstammungs- u. Verebungslehre* **42,** 237.
Metzner, R. (1894). *Arch. Physiol.* (*Physiol. Abt. Arch. Anat. Physiol.*), p. 309.
Meves, F. (1903). *Arch. Mikroskop. Anat.* **61,** 1.
Migeon, B. R., and Miller, C. S. (1968). *Science* **162,** 1005.
Minouchi, O. (1936). *Z. Zellforsch. Mikroskop. Anat.* **24,** 85.
Mizukami, I., and Gall, J. (1966). *J. Cell Biol.* **29,** 97.
Moens, P. B. (1968). *Chromosoma* **23,** 418.

Moens, P. B. (1969). *Chromosoma* **28,** 1.
Mohri, H. (1968). *Nature* **217,** 1053.
Molè-Bajer, J. (1958). *Chromosoma* **9,** 332.
Molè-Bajer, J. (1965). *J. Cell Biol.* **25,** 79.
Molè-Bajer, J. (1967). *Chromosoma* **22,** 465.
Molè-Bajer, J. (1969). *Chromosoma* **26,** 427.
Moor, H. (1967). *Protoplasma* **64,** 89.
Morgenthaler, H. (1948). *Rev. Suisse Zool.* **55,** 310.
Morowitz, H. J. (1966). *In* "Principles of Biomolecular Organization (G. E. W. Wolstenholme and M. O'Connor, eds.), pp. 446–462. Churchill, London.
Mortimer, R. K., and Hawthorne, D. C. (1966). *Genetics* **53,** 165.
Moses, M. J. (1968). *Ann. Rev. Genet.* **2,** 363.
Motta, J. J. (1967). *Mycologia* **59,** 370.
Motta, J. J. (1969). *Mycologia* **61,** 873.
Mottier, D. M. (1897). *Jahrb. Wiss. Botan.* **31,** 125.
Munkres, K. D., and Richards, F. M. (1965). *Arch. Biochim. Biophys.* **109,** 457.
Munkres, K. D., Giles, N. H., and Case, M. E. (1965). *Arch. Biochem. Biophys.* **109,** 397.
Murayama, M. (1966). *Science* **153,** 145.
Murray, R. G., Murray, A. S., and Pizzo, A. (1965). *J. Cell Biol.* **26,** 601.
Nagai, R., and Rebhun, L. I. (1966). *J. Ultrastruct. Res.* **14,** 571.
Nagano, T. (1966). *In* Fawcett, D. W. "An Atlas of Fine Structure," p. 54. Saunders, Philadelphia, Pennsylvania.
Nance, W. E. (1964). *Cold Spring Harbor Symp. Quant. Biol.* **29,** 415.
Nanney, D. L. (1964). *In* "The Role of Chromosomes in Development" (M. Locke, ed.), pp. 253–273. Academic Press, New York.
Nanney, D. L. (1968). *Science* **160,** 496.
Nanney, D. L., and Nagel, M. J. (1964). *J. Protozool.* **11,** 465.
Naville, A., and de Beaumont, J. (1933). *Arch. Anat. Microscop.* **29,** 199.
Nawaschin, S. (1927). *Ber. Deut. Botan. Ges.* **45,** 415.
Nebel, B. R., and Coulon, E. M. (1962). *Chromosoma* **13,** 272.
Newcomb, E. H., and Bonnett, H. T., Jr. (1965). *J. Cell Biol.* **27,** 575.
Newman, L. J. (1967). *Chromosoma* **22,** 258.
Nicklas, R. B. (1959). *Chromosoma* **10,** 301.
Nicklas, R. B. (1961). *Chromosoma* **12,** 97.
Nicklas, R. B. (1963). *Chromosoma* **14,** 276.
Nicklas, R. B. (1965). *J. Cell Biol.* **25,** 119.
Nicklas, R. B. (1966). *6th Ann. Meeting Am. Soc. Cell Biology, Houston, Texas.* Unpublished observations.
Nicklas, R. B. (1967). *Chromosoma* **21,** 17.
Nicklas, R. B., and Koch, C. (1969). *J. Cell Biol.* **43,** 40.
Nicklas, R. B., and Staehly, C. A. (1967). *Chromosoma* **21,** 1.
Novitski, E. (1952). *Genetics* **37,** 270.
Novitski, E. (1955). *J. Cellular Comp. Physiol.* **45** (Suppl. 2), 151.
Nur, U. (1963). *Chromosoma* **14,** 123.
Oftebro, R., and Wolf, I. (1968). *Exptl. Cell Res.* **48,** 39.
Ohnuki, Y. (1968). *Chromosoma* **25,** 402.

Östergren, G. (1950). *Hereditas* **36,** 1.
Östergren, G. (1951). *Hereditas* **37,** 85.
Östergren, G. (1961). *Chromosoma* **12,** 80.
Östergren, G., and Bajer, A. (1961). *Chromosoma* **12,** 72.
Östergren, G., Molè-Bajer, J., and Bajer, A. (1960). *Ann. N.Y. Acad. Sci.* **90,** 381.
Östergren, G., and Vigfusson, E. (1953). *Hereditas* **39,** 33.
Osterhout, W. J. V. (1902). *Proc. Calif. Acad. Sci.* [3] **2,** 255.
Outka, D. E., and Kluss, B. C. (1967). *J. Cell Biol.* **35,** 323.
Painter, T. S. (1918). *J. Exptl. Zool.* **24,** 445.
Parchman, L. G., and Stern, H. (1969). *Chromosoma* **26,** 298.
Pawaletz, N. (1967). *Naturwissenschaften* **54,** 533.
Payne, F. (1927). *J. Morphol.* **43,** 299.
Peacock, W. J. (1965). *Genetics* **51,** 573.
Peacock, W. J., and Erickson, J. (1964). *Drosophila Inform. Serv.* **39,** 107.
Peacock, W. J., and Erickson, J. (1965). *Genetics* **51,** 313.
Pease, D. C. (1946). *Biol. Bull.* **91,** 145.
Pease, D. C. (1963). *J. Cell Biol.* **18,** 313.
Pera, F., and Schwarzacher, H. G. (1969). *Chromosoma* **26,** 337.
Perkins, F. O., and Amon, J. P. (1969). *J. Protozool.* **16,** 235.
Peters, A., and Vaughn, J. E. (1967). *J. Cell Biol.* **32,** 113.
Peters, A., Proskauer, C. C., and Kaiserman-Abramof, I. R. (1968). *J. Cell Biol.* **39,** 604.
Phillips, D. M. (1966). *J. Cell Biol.* **30,** 477.
Phillips, D. M. (1967). *J. Cell Biol.* **33,** 73.
Pickett-Heaps, J. D. (1967). *Australian J. Biol. Sci.* **20,** 883.
Pickett-Heaps, J. D., and Fowke, L. C. (1969). *Australian J. Biol. Sci.* **22,** 857.
Pickett-Heaps, J. D., and Northcote, D. H. (1966). *J. Cell Sci.* **1,** 121.
Pontefract, R. D., Bergeron, G., and Thatcher, F. S. (1969). *J. Bacteriol.* **97,** 367.
Porter, K. R. (1966). *In* "Principles of Biomolecular Organization" (G. E. W. Wolstenholme and M. O'Connor, eds.), pp. 308–356. Little, Brown, Boston, Massachusetts.
Porter, K. R., and Machado, R. D. (1960). *J. Biophys. Biochem. Cytol.* **7,** 167.
Postgate, J. R., Crumpton, J. E., and Hunter, J. R. (1961). *J. Gen. Microbiol.* **24,** 15.
Powell, E. O. (1958). *J. Gen. Microbiol.* **18,** 382.
Prokofieva, A. (1935). *Cytologia* **6,** 148.
Ptashne, M. (1960). *Genetics* **45,** 499.
Puck, T. T., and Marcus, P. I. (1955). *Proc. Natl. Acad. Sci. U.S.* **41,** 432.
Pyne, C. K. (1968). *Compt. Rend. Acad. Sci. Ser. D* **267,** 755.
Rana, R. S. (1967). *Experientia* **23,** 119.
Randall, J., and Disbrey, C. (1965). *Proc. Roy. Soc.* **B162,** 473.
Raply, S., Robson, E. B., Harris, H., and Smith, S. M. (1967). *Ann. Human Genet.* **31,** 237.
Rashevsky, N. (1941). *Bull. Math. Biophys.* **3,** 1.
Raven, C. P. (1966). "An Outline of Developmental Physiology," 3rd ed. Pergamon, Oxford.
Ray-Chaudhuri, S. P., and Manna, G. K. (1952). *J. Genet.* **51,** 191.

Rebhun, L. I. (1959). *Biol. Bull.* **117,** 518.
Rebhun, L. I. (1960). *Ann. N.Y. Acad. Sci.* **90,** 357.
Rebhun, L. I. (1963). *In* "The Cell in Mitosis" (L. Levine, ed.), pp. 67–102. Academic Press, New York.
Rebhun, L. I. (1964). *In* "Primitive Motile Systems in Cell Biology" (R. D. Allen and N. Kamiya, eds.), pp. 503–525. Academic Press, New York.
Rebhun, L. I. (1967). *J. Gen. Physiol.* **50,** 223.
Rebhun, L. I., and Sander, G. (1967). *J. Cell Biol.* **34,** 859.
Rebhun, L. I., and Sawada, N. (1969). *Protoplasma* **68,** 1.
Renaud, F. L., Rowe, A. J., and Gibbons, I. R. (1968). *J. Cell Biol.* **36,** 79.
Rhea, R. P. (1966). *J. Ultrastruct. Res.* **15,** 349.
Rhoades, M. (1942). *Genetics* **27,** 395.
Rhoades, M. (1952). *In* "Heterosis" (J. W. Gowen, ed.), pp. 66–81. Iowa State College Press, Ames, Iowa.
Richardson, M. M. (1936). *J. Genet.* **32,** 411.
Richart, R. M., and Corfman, P. A. (1964). *Science* **144,** 65.
Ris, H. (1949). *Biol. Bull.* **96,** 90.
Ritossa, F. M., and Spiegelman, S. (1965). *Proc. Natl. Acad. Sci. U.S.* **53,** 737.
Ritossa, F. M., Atwood, K. C., Lindsley, D. L., and Spiegelman, S. (1965). *Natl. Cancer Inst. Monograph* **23,** 449.
Ritossa, F. M., Atwood, K. C., and Spiegelman, S. (1966). *Genetics* **54,** 663.
Roberts, H. S., and Johnson, N. S. (1956). *Biol. Bull.* **110,** 334.
Robinow, C. F., and Bakerspigel, A. (1965). *In* "The Fungi." (G. C. Ainsworth and A. E. Sussman, eds.), Vol. I, pp. 119–142. Academic Press, New York.
Robinow, C. F., and Marak, J. (1966). *J. Cell Biol.* **29,** 129.
Robbins, E., and Gonatas, N. K. (1964). *J. Cell Biol.* **21,** 429.
Robbins, E., and Jentzsch, G. (1969). *J. Cell Biol.* **40,** 678.
Robbins, E., Jentzsch, G., and Micali, A. (1968). *J. Cell Biol.* **36,** 329.
Rondanelli, E. G., Trenta, A., Magliulo, E., Vannini, V., Gerna, G., and Carosi, G. (1967). *Exptl. Cell Res.* **47,** 222.
Rosenbaum, J. L., and Child, F. M. (1967). *J. Cell Biol.* **34,** 345.
Rosenbaum, J. L., Moulder, J. E., and Ringo, D. L. (1969). *J. Cell Biol.* **41,** 600.
Rosenberger, R. F., and Kessel, M. (1968). *J. Bacteriol.* **96,** 1208.
Roth, L. E. (1967). *J. Cell Biol.* **34,** 47.
Roth, L. E., and Daniels, E. W. (1962). *J. Cell Biol.* **12,** 57.
Roth, L. E., and Shigenaka, Y. (1964). *J. Cell Biol.* **20,** 249.
Roth, L. E., Obetz, S. W., and Daniels, E. W. (1960). *J. Biophys. Biochem. Cytol.* **8,** 207.
Roth, L. E., Wilson, H. J., and Chakraborty, J. (1966). *J. Ultrastruct. Res.* **14,** 460.
Roth, T. F. (1966). *Protoplasma* **61,** 346.
Rudzinska, M. A. (1965). *J. Cell Biol.* **25,** 459.
Russell, L. B. (1964). *In* "The Role of Chromosomes in Development" (M. Locke, ed.), pp. 153–181. Academic Press, New York.
Ryter, A. (1968). *Bacteriol. Rev.* **32,** 39.
Ryter, A., and Jacob, F. (1964). *Ann. Inst. Pasteur* **107,** 384.
Ryter, A., and Jacob, F. (1966). *Ann. Inst. Pasteur* **110,** 801.
Sabnis, D. D., and Jacobs, W. P. (1967). *J. Cell Sci.* **2,** 465.

Sakai, H. (1966). *Biochim. Biophys. Acta* **112,** 132.
Sandler, L., Hiraizumi, Y., and Sandler, I. (1959). *Genetics* **44,** 233.
Satir, P. (1967). *J. Gen. Physiol.* **50,** 241.
Satir, P. (1968). *J. Cell Biol.* **39, 77**.
Sauaia, H., and Mazia, D. (1961). *Pathol. Biol.* **9,** 473.
Schmid, W. (1966). *Exptl. Cell Res.* **42,** 201.
Schmitt, F. O. (1968). *Proc. Natl. Acad. Sci. U.S.* **60,** 1092.
Schrader, F. (1936). *Biol. Bull.* **70,** 484.
Schrader, F. (1939). *Chromosoma* **1,** 230.
Schrader, F. (1946a). *Biol. Bull.* **90,** 19.
Schrader, F. (1946b). *Biol. Bull.* **90,** 265.
Schrader, F. (1947). *Chromosoma* **3,** 22.
Schrader, F. (1953). "Mitosis: The Movements of Chromosomes in Division," 2nd ed. Columbia Univ. Press, New York.
Schrantz, J.-P. (1967). *Compt. Rend. Acad. Sci. Ser. D* **264,** 1274.
Schreil, W. -H., and Christensen, R. J. (1968). *Proc. Natl. Acad. Sci. U.S.* **59,** 1152.
Schreiner, A., and Schreiner, K. E. (1905). *Archiv. Biol.* **21,** 315.
Schuel, H. (1961). *Biol. Bull* **120,** 384.
Schuster, F. (1963). *J. Protozool.* **10,** 297.
Schuster, F. (1964). *J. Protozool.* **11,** 207.
Schuster, F. (1965). *Protistologica* **1,** 49.
Schwarz, U., Asmus, A., and Frank, H. (1969). *J. Mol. Biol.* **41,** 419.
Schwarzacher, H. G., and Schnedl, W. (1966). *Nature* **209,** 107.
Schweiger, H. G., Master, R. W. P., and Werz, G. (1967). *Nature* **216,** 554.
Scott, A. C. (1936). *J. Morphol.* **59,** 485.
Scriba, M. E. L. (1967). *Arch. Entwicklungsmech. Organ.* **159,** 314.
Scriba, M. E. L. (1969). *Develop. Biol.* **19,** 160.
Sears, E. R. (1952). *Chromosoma* **4,** 535.
Seto, T., Kezer, J., and Pomerat, C. M. (1969). *Z. Zellforsch. Mikroskop. Anat.* **94,** 407.
Sherman, F., Stewart, J. W., Margoliash, E., Parker, J., and Campbell, W. (1966). *Proc. Natl. Acad. Sci. U.S.* **55,** 1498.
Sherman, F., Stewart, J. W., Parker, J. H., Inhaber, E., Shipman, N., Putterman, G. J., Gardisky, R. L., and Margoliash, E. (1968). *J. Biol. Chem.* **243,** 5446.
Shows, T. B., Ruddle, F. H., and Roderick, T. H. (1969). *Biochem. Genet.* **3,** 25.
Sisken, J. E., and Kinoshita, R. (1961). *Exptl. Cell Res.* **22,** 521.
Smith, D. W., and Hanawalt, P. C. (1967). *Biochim. Biophys. Acta* **149,** 519.
Smith, L. (1942). *Am. J. Botany* **29,** 451.
Smith, S. G. (1953). *Chromosoma* **5,** 555.
Smith-Sonneborn, J., and Plaut, W. (1967). *J. Cell Sci.* **2,** 225.
Snyder, L. A. (1961). *Cytologia* **26,** 50.
Snyder, R. W., and Young, F. E. (1969). *Biochem. Biophys. Res. Commun.* **35,** 354.
Somers, C. E., and Hsu, T. C. (1962). *Proc. Natl. Acad. Sci. U.S.* **48,** 937.
Sonneborn, T. M. (1963). *In* "The Nature of Biological Diversity" (J. M. Allen, ed.), pp. 165–219. McGraw-Hill, New York.
Sorokin, S. P. (1968). *J. Cell Sci.* **3,** 207.
Southern, D. I. (1969). *Chromosoma* **26,** 140.

Soyer, M. -O. (1969). *Compt. Rend. Acad. Sci. Ser. D* **268,** 2082.
Specht, W. (1961). *Z. Anat. Entwicklungsgeschichte* **122,** 266.
Staiger, H. (1954). *Chromosoma* **6,** 419.
Staiger, H. (1955). *Chromosoma* **7,** 181.
Stephens, R. E. (1968). *J. Mol. Biol.* **32,** 277.
Stich, H. (1954). *Chromosoma* **6,** 199.
Stone, G. E., Miller, O. L., Jr., and Prescott, D. M. (1965). *J. Cell Biol.* **25,** 171.
Storch, O. (1924). *Zool. Jahrb. (Abt. Anat. Ontongenie Tiere)* **45,** 309.
Strasburger, E. (1895). *Jahr. Wiss. Botan.* **28,** 151.
Stubblefield, E., and Brinkley, B. R. (1967). *Symp. Intern. Soc. Cell Biol.* **6,** 175–218.
Subirana, J. A. (1968). *J. Theoret. Biol.* **20,** 117.
Swann, M. M. (1952). *In* Hughes, A., "The Mitotic Cycle," pp. 119–133. Academic Press, New York.
Swanson, C. P., Merz, T., and Young, W. J. (1967). "Cytogenetics." Prentice-Hall, Englewood Cliffs, New Jersey.
Szollosi, D. (1964). *J. Cell Biol.* **21,** 465.
Tait, A. (1968). *Nature* **219,** 941.
Tauro, P., Halvorson, H. O., and Epstein, R. L. (1968). *Proc. Natl. Acad. Sci. U.S.* **59,** 277.
Taylor, A. I. (1968). *Nature* **219,** 1028.
Taylor, E. W. (1959). *J. Biophys. Biochem. Cytol.* **6,** 193.
Taylor, E. W. (1965). *J. Cell Biol.* **25,** 145.
Taylor, J. H. (1963). *In* "Molecular Genetics" (J. H. Taylor, ed.), pp. 65–111. Academic Press, New York.
Taylor, J. H., Haut, W. F., and Tung, J. (1962). *Proc. Natl. Acad. Sci. U.S.* **48,** 190.
Teplitz, R. L., Gustafson, P. E., and Pellett, O. L. (1968). *Exptl. Cell Res.* **52,** 379.
Terzakis, J. A., Sprinz, H., and Ward, R. A. (1967). *J. Cell Biol.* **34,** 311.
Tilney, L. G. (1968). *J. Cell Sci.* **3,** 549.
Tilney, L. G., and Byers, B. (1969). *J. Cell Biol.* **43,** 148.
Tilney, L. G., and Porter, K. R. (1965). *Protoplasma* **60,** 317.
Tilney, L. G., and Porter, K. R. (1967). *J. Cell Biol.* **34,** 327.
Ting, Y. C. (1966). *Cytologia* **31,** 324.
Tjio, J. H., and Levan, A. (1950a). *Anales Estac. Exptl. Aula Dei* **2,** 21.
Tjio, J. H., and Levan, A. (1950b). *Nature* **165,** 368.
Trankowsky, D. A. (1939). *Z. Zellforsch. Mikroskop. Anat.* **10,** 736.
Trosko, J. E., and Wolff, S. (1965). *J. Cell Biol.* **26,** 125.
Tucker, J. B. (1967). *J. Cell Sci.* **2,** 281.
Turner, F. R. (1968). *J. Cell Biol.* **37,** 370.
Ullerich, F. -H., Bauer, H., and Dietz, R. (1964). *Chromosoma* **15,** 591.
Underbrink, A. G., Sparrow, A. H., Rogers, A. F., and Pond, V. (1967). *Cytologia* **32,** 489.
Upcott, M. (1937). *Proc. Roy. Soc.* **B124,** 336.
Valentine, R. C., and Bradfield, J. R. G. (1954). *J. Gen. Microbiol.* **11,** 349.
Van Assel, S., and Brachet, J. (1966). *J. Embryol. Exptl. Morphol.* **15,** 143.
Virkki, N. (1967). *Hereditas* **57,** 275.
Virkki, N. (1968). *Caryologia* **21,** 47.
Vivier, E. (1965). *J. Microscop.* **4,** 559.

von Korff, K. (1901). *Anat. Anz.* **19,** 490.
Wagenaar, E. B. (1961a). *Can. J. Botany* **39,** 81.
Wagenaar, E. B. (1961b). *Can. J. Genet. Cytol.* **3,** 204.
Wagenaar, E. B. (1969). *Chromosoma* **26,** 410.
Walen, K. H. (1965). *Genetics* **51,** 915.
Wallace, H., and Birnstiel, M. L. (1966). *Biochim. Biophys. Acta* **114,** 296.
Walters, M. S. (1958). *Am. J. Botany* **45,** 271.
Weber, D. F. (1969). *Chromosoma* **27,** 354.
Went, H. A. (1959). *J. Biophys. Biochem. Cytol.* **6,** 447.
Went, H. A. (1966a). *Protoplasmatologia* **VI G1.**
Went, H. A. (1966b). *J. Cell Biol.* **30,** 555.
Went, H. A., and Mazia, D. (1959). *Exptl. Cell Res. Suppl.* **7,** 200.
Wettstein, R., and Sotelo, J. R. (1965). *J. Ultrastruct. Res.* **13,** 367.
White, M. J. D. (1946). *J. Exptl. Zool.* **102,** 179.
White, M. J. D. (1961). *Australian J. Zool.* **9,** 784.
Whitehouse, H. L. K. (1965). "Towards an Understanding of the Mechanism of Heredity." St. Martin's Press, New York.
Whitten, M. J. (1968). *Chromosoma* **24,** 37.
Wikswo, M. A., and Novales, R. N. (1969). *Biol. Bull.* **137,** 228.
Wilson, E. B. (1901a). *Arch. Entwicklungsmech. Organ.* **12,** 529.
Wilson, E. B. (1901b). *Arch. Entwicklungsmech. Organ.* **13,** 353.
Wilson, E. B. (1925). "The Cell in Development and Heredity," 3rd ed. Macmillan, New York.
Wilson, G. B., and Cheng, K. C. (1949). *J. Heredity* **40,** 2.
Wilson, H. J. (1967). *J. Cell Biol.* **35,** 145A.
Wilson, H. J. (1968). *Planta* **78,** 379.
Wilson, H. J. (1969). *J. Cell Biol.* **40,** 854.
Wilson, J. Y. (1956). *Genetica* **28,** 223.
Wohlfarth-Bottermann, K. E. (1964a). *In* "Primitive Motile Systems in Cell Biology" (R. D. Allen and N. Kamiya, eds.), pp. 79–109. Academic Press, New York.
Wohlfarth-Bottermann, K. E. (1964b). *Intern. Rev. Cytol.* **16,** 61.
Wohlman, A., and Allen, R. D. (1968). *J. Cell Sci.* **3,** 105.
Wolfe, S. L. (1965). *J. Cell Biol.* **25,** 408.
Wolfe, S. L., and Martin, P. G. (1968). *Exptl. Cell Res.* **50,** 140.
Woods, P. S. (1965). *Natl. Cancer Inst. Monograph* **23,** 271.
Woods, P. S., and Zubay, G. (1965). *Proc. Natl. Acad. Sci. U.S.* **54,** 1705.
Yamanaka, T., and Okada, Y. (1968). *Exptl. Cell Res.* **49,** 461.
Yamasaki, N. (1961). *Chromosoma* **11,** 479.
Yanders, A. F., Brewen, J. G., Peacock, W. J., and Goodchild, D. J. (1968). *Genetics* **59,** 245.
Yatsu, N. (1905). *J. Exptl. Zool.* **2,** 287.
Yerganian, G., and Nell, M. B. (1966). *Proc. Natl. Acad. Sci. U.S.* **55,** 1066.
Yoshida, A. (1967). *Proc. Natl. Acad. Sci. U.S.* **57,** 835.
Young, W. J. (1966). *J. Heredity* **57,** 58.
Young, W. J., Porter, J. E., and Childs, B. (1964). *Science* **143,** 140.

Zamboni, L., and Mastroianni, L., Jr. (1966). *J. Ultrastruct. Res.* **14,** 118.
Zickler, D. (1969). *Compt. Rend. Acad. Sci. Ser. D* **268,** 3040.
Ziegler, H. E. (1898). *Arch. Entwicklungsmech. Organ.* **6,** 249.
Zimmering, S., and Fowler, G. L. (1968). *Genet. Res.* **12,** 359.
Zimmerman, A. M., and Marsland, D. (1964). *Exptl. Cell Res.* **35,** 293.

AUTHOR INDEX

Numbers in italics refer to the pages on which the complete references are listed.

I

J

K

N

O

P

T

U

V

SUBJECT INDEX

M

S